ENDANGERED SEA LIFE !

For a free color catalog describing Gareth Stevens' list of high-quality books, call 1-800-542-2595 (USA) or 1-800-461-9120 (Canada). Gareth Stevens' Fax: (414) 225-0377.

Library of Congress Cataloging-in-Publication Data available upon request from publisher. Fax: (414) 225-0377 for the attention of the Publishing Records Department.

ISBN 0-8368-1425-8

Exclusive publication in North America in 1996 by
Gareth Stevens Publishing
1555 North RiverCenter Drive, Suite 201
Milwaukee, Wisconsin 53212, USA

A LOVELL JOHNS PRODUCTION created, designed, and produced by Lovell Johns, Ltd., 10 Hanborough Business Park, Long Hanborough, Witney, Oxfordshire OX8 8LH, UK.

U.S. series editor: Patricia Lantier-Sampon

Printed in the United Kingdom

1 2 3 4 5 6 7 8 9 99 98 97 96

ENDANGERED SEA LIFE !

Gareth Stevens Publishing
MILWAUKEE

CONTENTS

FOREWORD

Mark Collins, Director of the World Conservation Monitoring Centre.

In 1963, the IUCN Species Survival Commission, chaired by Sir Peter Scott, commissioned research and a series of books aimed at drawing to the attention of governments and the public the global threats to species. Sir Peter wanted more concerted action to address the problem of extinction. The first Red Data Book, published in 1969, was written by James Fisher, Noel Simon, and Jack Vincent. There had been earlier books that highlighted animals under threat and the possibility of extinction, the most important being written by G. M. Allen in 1942. The increasing threat to species and our knowledge of these threats has resulted in nearly 6,000 species being listed as threatened in the most recent IUCN Red List of Threatened Animals. (IUCN uses different categories of threatened species, of which the most crucial category is Endangered.)

Knowledge of the conservation status of species is required so priorities can be set and management actions taken to protect them. The original Red Data Books were global assessments of species. However, many of these globally threatened species are found in only one country, and it has become increasingly important for each country to assess its own species and decide which should be listed as Threatened. There are

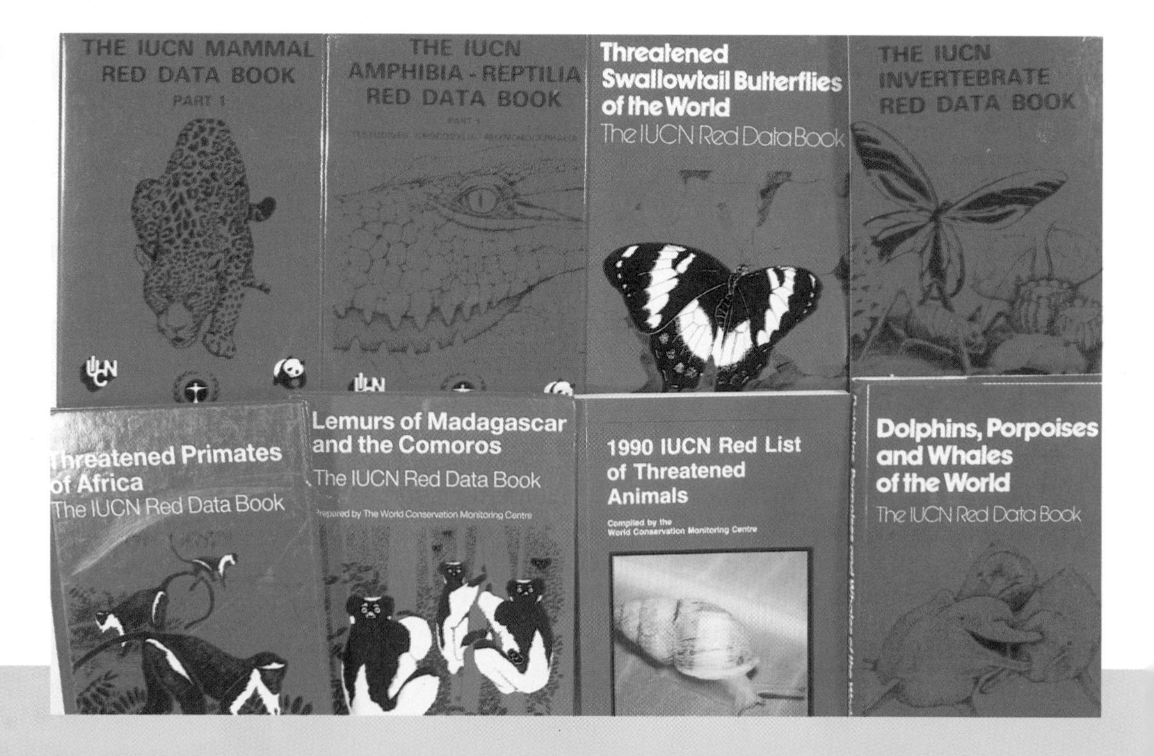

Some of the many Red Data Books published since the first one appeared in 1969.

now various National Red Data Books covering substantial areas of the world.

The very fact that there are so many threatened species makes it very difficult to publish books on their status and distribution and, for some of them, we do not have detailed information. This *Endangered!* series aims to provide sound knowledge of 150 selected endangered animals and their natural habitats to a wider audience, particularly young people.

Our knowledge of threatened species can only be as good as the research work that has been carried out on them and, as the charts on this page show, the conservation status of much of the world's wildlife has not yet been assessed. Even for mammals, only about 55% of the species have been assessed. The only major group of which all species have been assessed are birds, and yet there are still large gaps in our knowledge of the status and trends in bird population numbers. However, their attractiveness and the interest shown in them by a great many people have improved the information available. Marine fish, despite their importance as a valuable food source throughout the world, tend to be assessed for conservation purposes only when their populations reach such a low point that it is no longer viable to catch them commercially. The 1994 IUCN Red List of Threatened Animals lists 177 endangered mammals representing 3.8% of the total number of mammal species and 188 birds representing 1.9% of the total number of species. Information on birds is compiled by BirdLife International.

The importance of identifying threatened species cannot be stressed enough. There have been many cases where conservation action has been taken as a result of the listing of species as endangered. The vicuna, a camel-like animal that lives in the high Andes of South America whose wool is said to be the finest in the world, was extremely abundant in ancient times but has been over-exploited since the European colonization of South America. By 1965, it was reduced to only 6,000 animals.

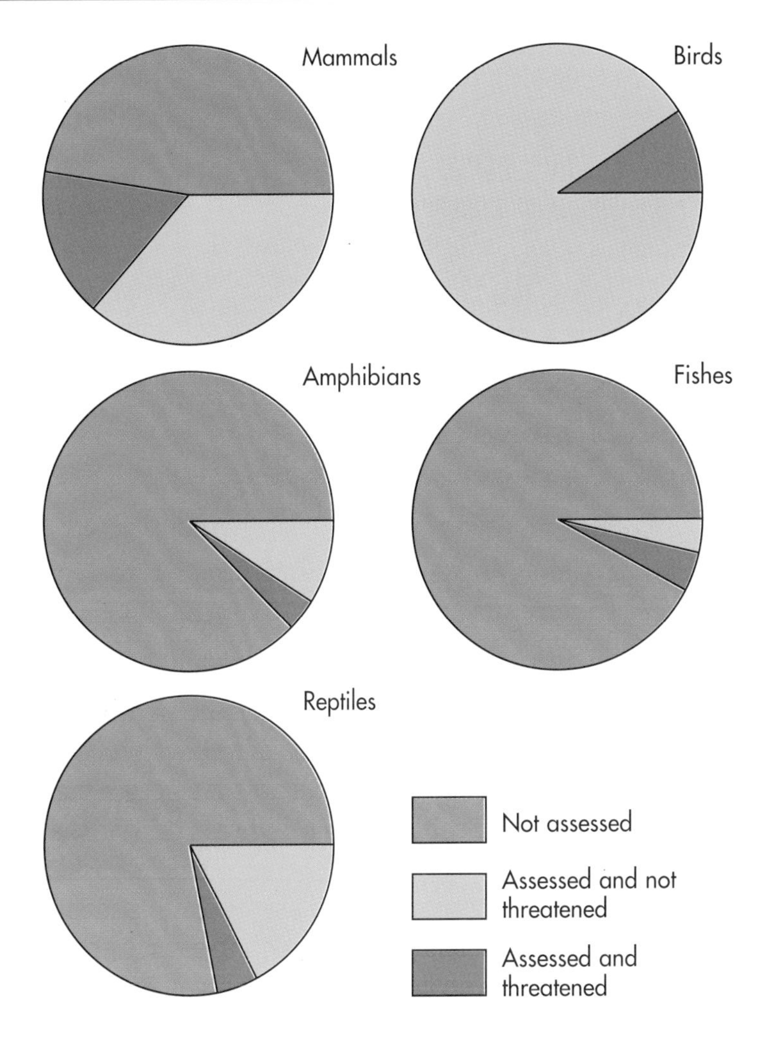

By protecting the vicuna from hunting and by establishing reserves, the population has steadily increased and is now in the region of 160,000. The vicuna is no longer an endangered species but is listed as Vulnerable. One high-profile endangered animal is the Indian Tiger, whose numbers had dropped to fewer than 2,000 in India when the first census was taken in 1972. Urgent conservation measures were taken, reserves were set up, and a great deal of expertise in their management has resulted in a population increase to its current level of about 3,250. Other measures included the halting of trade in tiger skins and other products such as bones and blood used in eastern traditional medicines. However, the other subspecies of tigers have not had this same protection, and their numbers are dwindling day by

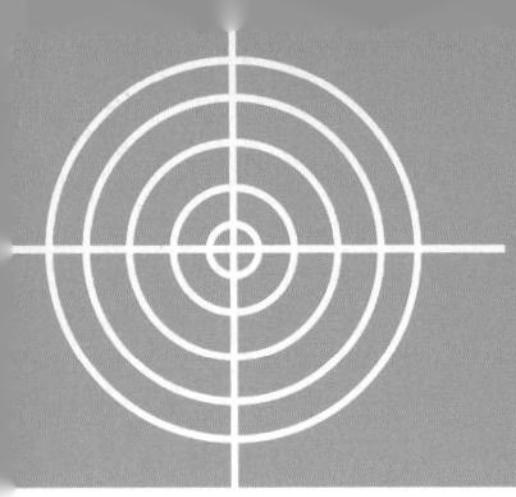

day. Gray whales were also endangered. They migrate down the west coast of North America from arctic waters to the coast of Mexico and southern California to mate, returning for the rest of the year to feed and give birth to their young. As their migration route was so well known, hunting was easy, and, as a result, their numbers had dropped to only a few hundred. Since hunting control measures began, the number of gray whales is now in excess of 21,000, and they are no longer listed as endangered.

WCMC

The World Conservation Monitoring Centre in Cambridge, in the United Kingdom, has been the focal point of the management and integration of information on endangered plant and animal species for more than fifteen years. WCMC's databases also cover the trade in wildlife throughout the world, information on the importance and number of areas set up to protect the world's wildlife, and a Biodiversity Map Library that holds mapped data on many of the world's important sites and ecosystems. It was IUCN, through its Species Survival Commission, that first established the World Conservation Monitoring Centre as its information database for species and ecosystems throughout the world. WCMC now carries on this role with the support of two other partners: the World Wide Fund For Nature and the United Nations Environment Programme.

IUCN — The World Conservation Union

Founded in 1948, The World Conservation Union brings together states, government agencies, and a diverse range of nongovernmental organizations in a unique world partnership: over 800 members in all, spread across some 125 countries. As a Union, IUCN seeks to influence, encourage, and assist societies throughout the world to conserve the integrity and diversity of nature and to ensure that any use of natural resources is equitable and ecologically sustainable. The World Conservation Union builds on the strengths of its members, networks, and partners to enhance their capacity and to support global alliances to safeguard natural resources at local, regional, and global levels.

Various organizations too numerous to mention help countries protect their wildlife. We urge you to support these organizations so the list of endangered species does not continue to grow. Your voice will be added to the many millions that are urging international cooperation for the protection and wise use of the wildlife that is such an important part of our natural heritage.

Headquarters of the World Conservation Monitoring Centre, Cambridge, England.

MARINE OTTER

The marine otter is also sometimes known as the chingungo. It is similar to the river-dwelling otters of Europe, Asia, and North America but spends its life on the coast and hunts in the sea. The original range of the species was from Chimbote, in central Peru, through Chile, Cape Horn, Isla de Los Estados, and southern Argentina. The marine otter probably does not survive in Argentina, except perhaps near Isla de Los Estados. It is extinct in many other parts of its original range and rare in places where it does survive.

Marine otters live on rocky coasts and in sheltered bays. They need dense vegetation for shelter on land and good growths of seaweed for feeding. They hunt for shellfish and fish in the sea and for prawns in freshwater lagoons. Female otters have one or two cubs each year.

The marine otter became rare because it was hunted for its fur. It is now protected, but poaching continues because one otter skin is worth three months of regular wages for a fisherman. Marine otters are also shot by fishermen who think the otters compete with them for fish, shellfish, and prawns. Harvesting seaweed and overfishing by humans make life difficult for marine otters. Some are also drowned in fishing nets.

Although the marine otter is protected, its fur is very valuable, and poaching continues.

FACTS

SCIENTIFIC NAME: *Lutra felina*

RED DATA BOOK: Vulnerable

AVERAGE FULL-GROWN SIZE:
Total length 34-45 inches (87-115 centimeters)

FULL-GROWN WEIGHT: 8.8-9.9 pounds (4-4.5 kilograms)

LIFE SPAN: Not known

JUAN FERNANDEZ FUR SEAL

The Juan Fernandez fur seal lives only on the Juan Fernandez and Deventuradas islands off the western coast of Chile. The species was discovered in 1683. Four years later, sealers visited the Juan Fernandez Islands to hunt seals for their skins. Sealing continued for over one hundred years; during this time, a seal population of several million virtually disappeared. In the 1950s, experts thought the Juan Fernandez fur seal might be extinct. Then reports came of seals being seen on the islands. In 1965, two hundred fur seals, including some pups, were discovered. Since that year, numbers have increased, and there are now about twelve thousand.

The seals spend most of the year at the breeding grounds, and the single pups are born in summer. The seals' diet consists of fish, squid, and octopus.

The Juan Fernandez fur seals have been protected by Chilean law since 1978. A few are killed illegally, however, by fishermen for bait. More importantly, fishing around the islands may rob the fur seals of food. More research is necessary to determine exactly what the fur seals eat and how fishing can be controlled to help the seal population increase.

FACTS

SCIENTIFIC NAME: *Arctocephalus philippii*

RED DATA BOOK: Vulnerable

AVERAGE FULL-GROWN SIZE: Total length Males 4.9-6.6 feet (1.5-2 meters); females 4.6 feet (1.4 m)

FULL-GROWN WEIGHT: Males 308.7 pounds (140 kilograms); females 110 pounds (50 kg)

LIFE SPAN: Not known

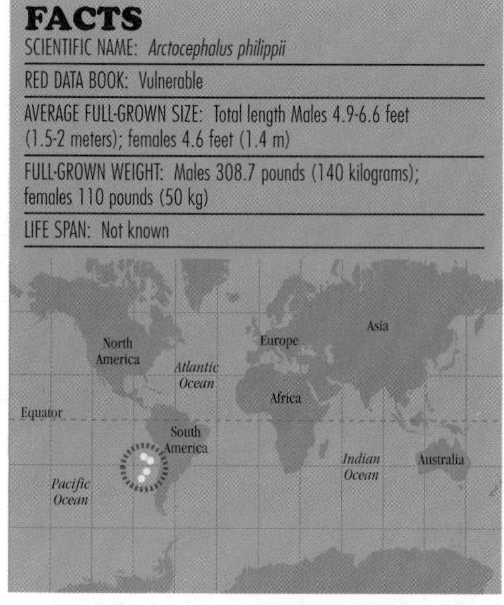

Fears that the Juan Fernandez fur seal might be extinct were fortunately unfounded. The world population has increased to approximately twelve thousand seals.

GUADALUPE FUR SEAL

The Guadalupe fur seal is found only on Isla Guadalupe, an island off the coast of Baja California, in Mexico. It was almost killed off completely by sealers before anyone realized it was different from any other species of fur seal. It may have once lived on other islands, including San Miguel off California in the United States, where Guadalupe fur seals are still occasionally seen.

Between 1800 and 1820, the Guadalupe fur seal was almost wiped out by sealers, but enough animals survived for hunting to continue until 1894. Thereafter, the species was sighted infrequently. Two were taken to San Diego Zoo in 1928, and a small breeding colony was rediscovered only in 1954.

Guadalupe fur seals have been seen eating fish and squid, but, other than this, scientists know very little about their habits. They live near their rocky breeding beaches all year round.

Laws in Mexico and the United States protect the Guadalupe fur seal. The island of Guadalupe was declared a seal reserve in 1975. The population is growing, and there are now about six thousand animals.

A Guadalupe fur seal rests on rocks and grooms itself.

FACTS

SCIENTIFIC NAME: *Arctocephalus townsendii*

RED DATA BOOK: Vulnerable

AVERAGE FULL-GROWN SIZE: Head and body length Males 6.6 feet (2 m); females 4.4 feet (1.35 m)

FULL-GROWN WEIGHT: Males 309 pounds (140 kg); females 110 pounds (50 kg)

LIFE SPAN: Not known

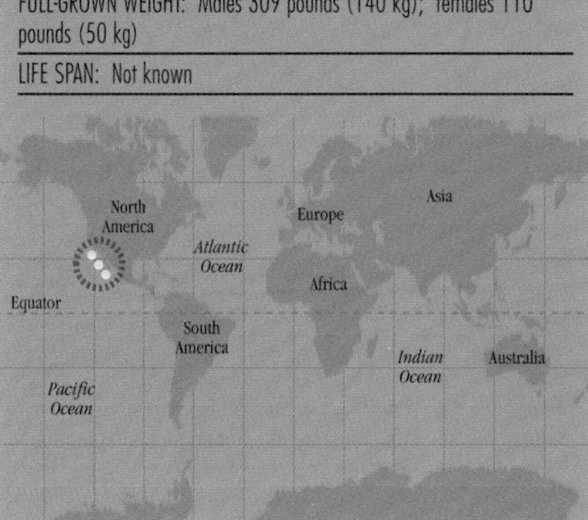

STELLER'S SEA LION

A large group of Alaskan Steller's sea lions rests on rocks.

FACTS
SCIENTIFIC NAME: *Eumetopias jubatus*
RED DATA BOOK: Vulnerable
AVERAGE FULL-GROWN SIZE: Head and body length Males 9 feet (2.8 m); females 7.5 feet (2.3 m)
FULL-GROWN WEIGHT: Males 2,205 pounds (1,000 kg); females 580 pounds (263 kg)
LIFE SPAN: Males 20 years; females 30 years

The largest of all the sea lions, Steller's sea lion, or the northern sea lion, lives only in the North Pacific. It lives from Hokkaido, Japan, northward around the eastern coast of Asia, across the Bering Sea, and down the western coast of America to San Miguel Island in the Californian Channel Islands. No Steller's sea lions have been seen in the Channel Islands since 1984, and most of the population is in the Aleutian Islands.

Scientists believe Steller's sea lions move quite a bit through the North Pacific for most of the year, but they return to rocky coasts and islands for the breeding season. The pup is born in midsummer, and it feeds on its mother's milk for one year or maybe longer. The adults feed mainly on fish and squid, but they also hunt fur seals, ringed seals, and sea otters.

Steller's sea lions have always been hunted by Aleuts for their skins and meat, but this did not seriously reduce the population. In 1978, experts estimated a population of as many as 240,000 to 300,000 Steller's sea lions. Since then, their numbers have dropped to about one-third of this amount. The decline has been greatest in the Aleutian Islands. The main reason is probably disease, but the sea lions may be starving as well. They do not have enough to eat because fishermen catch millions of tons of their food.

AUSTRALIAN SEA LION

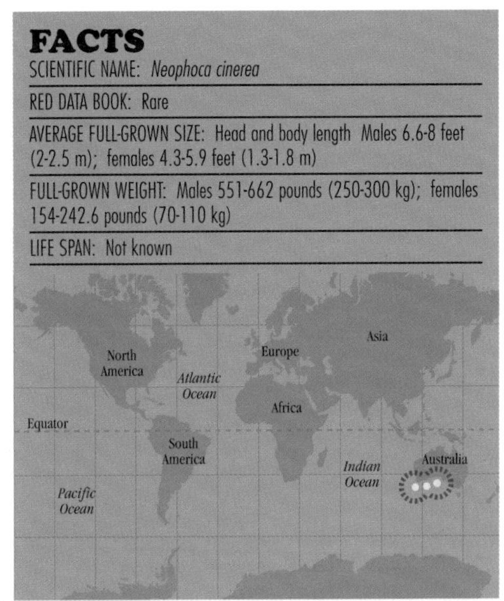

FACTS

SCIENTIFIC NAME: *Neophoca cinerea*

RED DATA BOOK: Rare

AVERAGE FULL-GROWN SIZE: Head and body length Males 6.6-8 feet (2-2.5 m); females 4.3-5.9 feet (1.3-1.8 m)

FULL-GROWN WEIGHT: Males 551-662 pounds (250-300 kg); females 154-242.6 pounds (70-110 kg)

LIFE SPAN: Not known

Before Europeans arrived in Australia, the Aborigines hunted the Australian sea lion for food. The original range of the species has been revealed by finding bones at old Aboriginal camp sites. The range used to extend around the southern coast of Australia from Houtman Abrolhos, north of Perth, to islands in the Bass Strait and along the coast of Tasmania. It now extends no farther than the Pages Islands, south of Adelaide.

The Australian sea lion lives on islands off the mainland of Australia, and not much is known about its habits. It eats fish, including small sharks and barracuda, crustaceans, and squid. Sometimes it catches and eats small penguins. Unlike other seals, the Australian sea lion does not have a regular breeding season, and pups are born at any time of the year.

Australian sea lions were hunted for their skins during the eighteenth and nineteenth centuries and became rare. They are now protected, and their numbers have risen to 10,000 to 12,000.

A few Australian sea lions are killed by fishermen because the sea lions compete for fish. Other sea lions die by drowning in fishing nets and crayfish pots. However, the sea lions have become a tourist attraction on Kangaroo Island near Adelaide. This may help preserve them.

The Australian sea lion's popularity with tourists should help preserve the species.

HOOKER'S SEA LION

Two Hooker's sea lion pups rest on the shore together.

Also known as the New Zealand sea lion, the Hooker's sea lion has a very restricted range. The main colonies thrive on islands to the south of New Zealand. These include the Auckland Islands, Snares Islands, and Campbell Island. Outside the breeding season, the sea lions swim out to sea, and they may come ashore on the southern edge of South Island, in New Zealand.

The Hooker's sea lions eat squid, crustaceans, and fish. They also occasionally eat penguins. The sea lions gather to breed at particular beaches each year, between October and January. Females give birth to a single pup.

When the islands south of New Zealand were first discovered, sealers soon arrived to slaughter the lions. They quickly decimated the population, but a few sea lions survived. The species was given protection in 1881. Some poaching continued, but numbers increased. The present population numbers 10,000 to 15,000 sea lions.

The current threat to the Hooker's sea lion comes from the fishing industry. Squid are caught within 31 miles (50 kilometers) of the breeding beaches, and 100 to 200 sea lions are caught accidentally every year. Other sea lions become trapped in trawling nets. The fishing problem can be solved if the ocean areas where the sea lions feed are declared sanctuaries, and fishing is forbidden. Another problem is that pups get stuck in the burrows dug by rabbits that have been introduced to the sea lions' breeding islands.

MEDITERRANEAN MONK SEAL

The Mediterranean monk seal is one of three true seals that live or once lived in warm waters. The Hawaiian monk seal is also endangered, and the Caribbean monk seal is extinct. The Mediterranean monk seal used to live throughout the Mediterranean and Black seas and along the Atlantic coast of northwestern Africa, including Madeira and the Canary Islands.

Monk seals eat a variety of fish and octopus, which are probably caught in shallow water. The single pup is born in a sheltered cave.

The increasing human population and activity around the Mediterranean Sea have caused the downfall of the monk seal. It is a shy animal that needs undisturbed coastal areas. Fishermen kill seals, which they see as competitors for fish, but the seals also suffer because fishermen have reduced the numbers of fish. Scuba divers also disturb the seals, especially during the breeding season.

The Mediterranean monk seal now lives only in a few parts of its former range. There are three major populations: two in the Aegean Sea, and one along the Atlantic coasts of southern Western Sahara and Mauritania. There is also a small colony at the Desertas Islands of Madeira. Scientists are not certain if any Mediterranean monk seals remain in the Black Sea. The total population is only a few hundred, of which 200 to 250 are near Greece, and numbers are declining.

Reserves have been set up on the Greek and Turkish coasts, and fishermen have been persuaded not to kill the seals but rather to take visitors to see them. A group of French scientists will attempt to breed monk seals in captivity when experts decide it is safe to move a group of wild seals into captivity.

Mediterranean monk seals are shy creatures that need undisturbed coastlines in order to thrive.

FACTS

SCIENTIFIC NAME: *Monachus monachus*

RED DATA BOOK: Endangered

AVERAGE FULL-GROWN SIZE:
Total length Males 7.9 feet (2.41 m); females 7.8 feet (2.38 m)

FULL-GROWN WEIGHT: Males 695 pounds (315 kg); females 662 pounds (300 kg)

LIFE SPAN: 23 years (in captivity)

HAWAIIAN MONK SEAL

The Hawaiian monk seal is found only in the Leeward Islands, the chain of small islands lying northwest of Hawaii. It has probably never lived anywhere else, but it is now confined to only a few of these islands.

Most of a monk seal's life is spent near the breeding beaches, but some seals travel to distant islands and spend time in the open sea. The seal's food is mainly octopus, lobster, and coral reef fish that are caught in shallow water. Monk seals do not gather in large groups to breed, as other seals do, but the females sometimes gather where there is shade from the tropical sun. The single pup stays with its mother for six weeks. Monk seals were once hunted for their flesh, skins, and blubber; as early as 1824, experts believed that a sealing ship had killed the last ones. However, some survived, and 1,500 more skins were taken in 1859. Even though sealing was banned at this point, the population continued to decline. It is difficult to count the seals accurately because they are scattered over many small islands, but their numbers dropped by 50 percent to about 700 in 1980. They have since increased to about 1,500.

The Hawaiian monk seal is protected, and the islands they inhabit are part of the Hawaiian Islands National Wildlife Refuge. However, commercial fishing is increasing in the area, and the seals may be robbed of their food or become tangled in nets.

The growth of the Hawaiian monk seal population is being helped by capturing pups that have been abandoned and feeding them until they can live on their own. Out of twenty pups released at Kure Atoll, fourteen are known to have reached maturity, and several have borne pups.

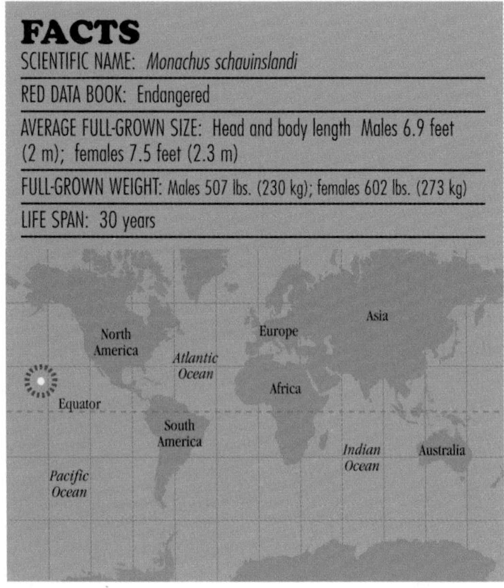

FACTS

SCIENTIFIC NAME: *Monachus schauinslandi*

RED DATA BOOK: Endangered

AVERAGE FULL-GROWN SIZE: Head and body length Males 6.9 feet (2 m); females 7.5 feet (2.3 m)

FULL-GROWN WEIGHT: Males 507 lbs. (230 kg); females 602 lbs. (273 kg)

LIFE SPAN: 30 years

Unlike other seals, Hawaiian monk seals do not gather in large groups to breed.

CASPIAN SEAL

Caspian seals have been hunted for over two hundred years, but scientists need to do more research to determine how many seals can be hunted without posing a threat to the survival of the species.

Twenty million years ago, the Tethys Sea covered western Asia, joining the Atlantic and Indian oceans. As mountain ranges formed in Europe and Asia, the area now known as the Caspian Sea was cut off, forming a landlocked sea. Since then, the population of seals in the sea has gradually evolved into a separate species known as the Caspian seals.

Caspian seals live throughout the sea, including river estuaries that flow into it. In winter, the seals prefer the northern end where the water freezes early, providing them with ice floes. The seals give birth to their pups there in January. In spring, the seals migrate to the southern end of the sea where the water is deeper and cooler. Caspian seals eat several kinds of fish and crustaceans.

An annual hunt for Caspian seals has existed for more than 200 years. At one time, 160,000 were killed every year. A quota of 50,000 per year out of a population of about half a million seals presently exists. Fishing also kills seals that become entangled in nets, and there is a growing threat from pollution of the Caspian Sea. Female seals are protected during the breeding season, but experts need to learn much more about the Caspian seal so that a safe number of seals can be hunted without endangering the species.

FACTS

SCIENTIFIC NAME: *Phoca caspica*

RED DATA BOOK: Vulnerable

AVERAGE FULL-GROWN SIZE: Head and body length Males 4.9 feet (1.5 m); females 4.6 feet (1.4 m)

FULL-GROWN WEIGHT: 154-176 pounds (70-80 kg)

LIFE SPAN: 35 years

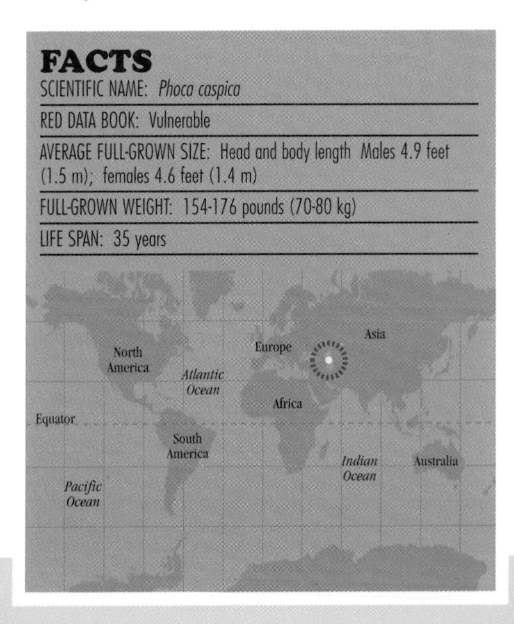

POLAR BEAR

The polar bear is closely related to the brown bear and has special characteristics suited to life in the Arctic. It is a sea mammal that spends most of its time floating on packs of ice. It is also a powerful swimmer.

Polar bears are found all around the Arctic. Most live near the coasts of North America, Europe, and Asia, as well as Greenland and Spitsbergen. Occasionally, one comes as far south as Iceland. The bears used to live in Newfoundland and the Gulf of St. Lawrence in Canada, but they have left these places because of growing human populations.

Polar bears eat seals that they hunt on the ice. They sometimes also kill narwhals and beluga whales. They will even eat lemmings, birds, and bird eggs if there is no other food available. The bears are normally solitary, but they may gather at good sources of food, such as dead whales and rubbish dumps. The females make a den in the snow for the winter, where they give birth to one to three cubs. The cubs stay with their mothers for two years, so females only give birth every three years.

Polar bears have always been important to hunting peoples, such as the Inuit, who live in the Arctic. Whalers, sealers, and travelers from regions to the south also hunted bears for sport and fur. But over-hunting has caused the numbers of polar bears to decline. In 1973, the five countries with polar bears in their territories agreed to control hunting. Local people are still allowed to hunt a certain number (150 per year in Greenland). The total number of polar bears has risen to 40,000, but scientists now realize that too many are still being killed, especially by poachers. Polar bears also face problems from disturbance and pollution as more people visit the Arctic for work or holidays.

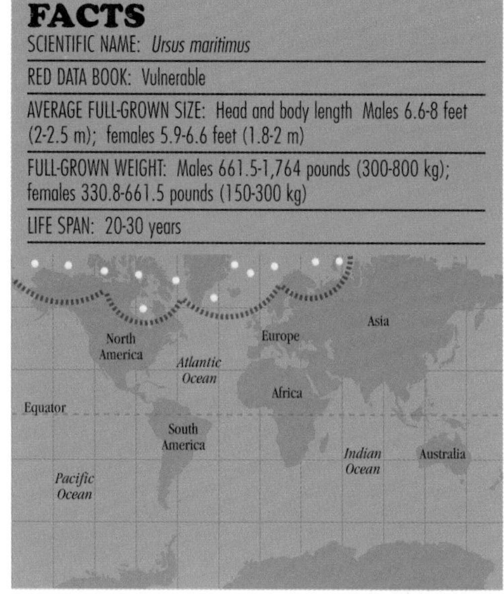

FACTS

SCIENTIFIC NAME: *Ursus maritimus*

RED DATA BOOK: Vulnerable

AVERAGE FULL-GROWN SIZE: Head and body length Males 6.6-8 feet (2-2.5 m); females 5.9-6.6 feet (1.8-2 m)

FULL-GROWN WEIGHT: Males 661.5-1,764 pounds (300-800 kg); females 330.8-661.5 pounds (150-300 kg)

LIFE SPAN: 20-30 years

This polar bear was photographed at Cape Churchill in Canada, where the bears have become a tourist attraction.

BOWHEAD WHALE

The bowhead whale, also known as the Greenland right whale, is confined to Arctic seas. There are four populations. In the Atlantic, one population lives around the islands of Spitsbergen, and the second between Canada and Greenland from Labrador northward. In the Pacific, bowheads live in the Okhotsk Sea and in the Bering Sea. Bowheads do not migrate as far as other whales, and those living in the Okhotsk Sea remain there all year.

The seas where bowheads live often freeze, but the whales survive by breaking the ice with their heads to breathe. They can crack ice 12 inches (30 cm) thick. For most of the year, bowheads live alone or in pairs, but small groups form during the breeding season and migration. There is usually a single calf, but there have been reports of twins.

Europeans began hunting bowheads around Spitsbergen in the early 1600s. Hunting followed in other areas, and by the twentieth century, bowheads had become very rare throughout their range. They are slow swimmers that are easy to catch, and they yield large quantities of oil and baleen (whalebone). The species is now protected except for small numbers caught by Eskimos in Alaska.

From a world population once estimated at about 56,000, there are now only 9,000 to 12,000 bowheads left. Although the Bering Sea population is around 7,500, there are less than 100 around Spitsbergen. Although the bowheads were considered near extinction at one time, improved methods of counting whales have proven otherwise.

FACTS

SCIENTIFIC NAME: *Balaena mysticetus*

RED DATA BOOK: Vulnerable

AVERAGE FULL-GROWN SIZE:
Head and body length 45.9-59 feet (14-18 m)

FULL-GROWN WEIGHT: 77 tons (70 tonnes)

LIFE SPAN: Not known

A bowhead whale photographed in the eastern Canadian Arctic.

SOUTHERN RIGHT WHALE

The right whales are so-called because they used to be the "right" whales to hunt. They swim slowly, so they are easy to harpoon from a small boat. They also yield plenty of oil and baleen (whalebone).

Southern right whales migrate between subantarctic seas, where they spend the summers, and temperate seas, where they spend the winters. They bear their calves in shallow, coastal waters around Australia and New Zealand, South America, and southern Africa. They used to travel in groups of one hundred, but now much smaller groups are common. They eat small crustaceans.

The southern right whale was hunted after the other species of right whales (bowhead and northern right whales) became rare. At the end of the eighteenth century, whalers from Europe and the United States sailed to the Southern Hemisphere to find new whale populations. After 60 to 70 years, so many southern right whales had been killed that they, too, were becoming rare. The species has been protected since 1935.

In 1972, scientists estimated 4,300 surviving southern right whales. Under protection, some populations have been increasing steadily at 7 percent each year, although females bear only one calf every two to five years. Groups of migrating right whales are becoming a familiar sight in some places where they swim close to the shore.

A southern right whale is shadowed by its calf in South African coastal waters.

FACTS

SCIENTIFIC NAME: *Eubalaena australis*

RED DATA BOOK: Vulnerable

AVERAGE FULL-GROWN SIZE:
Head and body length 32.8 feet (10 m)

FULL-GROWN WEIGHT: Males 48,510 pounds (22,000 kg); females 50,715 pounds (23,000 kg)

LIFE SPAN: Not known

North America
Atlantic Ocean
Europe
Asia
Africa
Equator
South America
Indian Ocean
Australia
Pacific Ocean

NORTHERN RIGHT WHALE

The northern right whale was the first whale to be hunted, and it is now the rarest of the great whales. The Japanese began hunting it in the tenth century, and the Spanish and French Basques in the eleventh century. The species remained the main target of the world's whaling industries until the seventeenth century.

Northern right whales live in the North Pacific and North Atlantic. They range from the Bering Sea, southern Greenland, and northern Norway in the north to Baja California, the Gulf of Mexico, and the Azores in the south. By keeping away from tropical waters, the species does not overlap with the very closely related southern right whale.

The food of the northern right whale is mainly small crustaceans, such as copepods, that live in vast swarms. Before they became so rare, the whales traveled in large groups. Calving takes place in shallow water, and the females give birth to one calf in alternate years.

Nothing is known about the size of the original populations of northern right whales. Present populations are near extinction, however, and there are probably only 100 to 200 in the Pacific Ocean and perhaps more in the Atlantic Ocean.

The species was given full protection in 1935, when it had already disappeared from many places. However, the former Soviet Union continued hunting northern right whales illegally. The remaining whales are threatened by collision with ships and drowning in fishing nets. Over one-third of whale deaths in the western North Atlantic are caused in these ways. These hazards make it very difficult for the populations to recover.

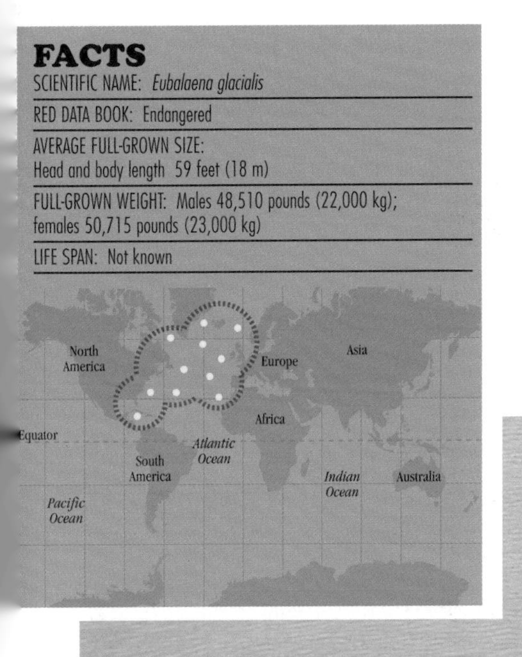

FACTS

SCIENTIFIC NAME: *Eubalaena glacialis*

RED DATA BOOK: Endangered

AVERAGE FULL-GROWN SIZE:
Head and body length 59 feet (18 m)

FULL-GROWN WEIGHT: Males 48,510 pounds (22,000 kg); females 50,715 pounds (23,000 kg)

LIFE SPAN: Not known

So few northern right whales remain that it will be very difficult for numbers to increase, making accidental deaths by drowning in fishing nets and collisions with ships even more serious.

SEI WHALE

An underwater close-up of the Sei whale clearly shows the pointed snout that characterizes all rorquals.

The sei whale is probably the fastest swimmer of all the giant rorqual whales, a group of whalebone whales that also includes fin, blue, and humpback whales. It can swim 12.5 to 15.5 miles (20 to 25 km) per hour when chased. It lives in almost all the world's oceans and seas, except for the coldest polar waters. Its main home is in temperate waters, and it stays in deep water. Only the largest individuals enter icy seas.

Sei whales live in small groups of three to five. They feed on a wide range of food, from the smallest copepod crustaceans to squid and shoaling fish. A single calf is born every two or three years.

Japanese whalers have caught sei whales since the seventeenth century. Most modern whalers hunted them only on a small scale until the larger blue, fin, and humpback whales became difficult to find. During the 1960s and 1970s, large numbers of sei whales were killed in the Southern Ocean (also referred to as either the Antarctic Ocean, or Southern Atlantic, Southern Pacific, or Southern Indian oceans), so they quickly became scarce. Hunting was illegal in the North Pacific by 1971, but the species was not given protection in the Southern Ocean until 1977. The whales received worldwide protection in 1979, except in the Denmark Strait, between Iceland and Greenland, where a few can be hunted each year.

The world population of sei whales is about 60,000 to 70,000. About 37,000 live in the Southern Ocean, compared with the original population of about 100,000.

FACTS

SCIENTIFIC NAME: *Balaenoptera borealis*

RED DATA BOOK: Vulnerable

AVERAGE FULL-GROWN SIZE: Head and body length Males 49 feet (15 m); females 52.5 feet (16 m)

FULL-GROWN WEIGHT: 33 tons (30 tonnes)

LIFE SPAN: 60 years

BLUE WHALE

Just over one hundred years ago, the harpoon cannon, which could be mounted on fast steamships, was invented. The combination of cannon and speed enabled whalers to catch the largest and fastest whales. These are the blue whale and its relatives, called rorquals.

The blue whale is probably the largest animal ever to have lived. A female measuring 110 feet (33.6 m) was the largest ever recorded. Large whales yielded more oil and were more valuable to whalers. For this reason, the blue whale was slaughtered more relentlessly than any other species by whalers.

Blue whales live in all the world's oceans, where they usually swim alone or in groups of two or three. They breed in warm tropical or subtropical waters, where the females give birth to a single calf every two or three years. They migrate to cold polar waters in the summer to feed on swarms of shrimplike crustaceans called krill. A blue whale eats between 2.2 to 4.4 tons (2-4 tonnes) of krill per day.

Before whaling started in the Southern Ocean in 1905, 150,000 to 200,000 blue whales visited antarctic waters each year. For the next forty years, they were slaughtered in large numbers, reaching a peak in 1930 when 30,000 blue whales were killed. By this time, scientists warned that the species was being overhunted. Blue whales were finally given full protection in 1966, when only a few hundred remained.

Worldwide, there are now several thousand blue whales, including nearly 4,000 in the northeastern Pacific and 10,000 to 12,000 in the Southern Hemisphere. The southern population is now only 5 percent of the original. Numbers are increasing in some places, such as the northeastern Atlantic, but total protection will be needed for many years for blue whales to become plentiful again.

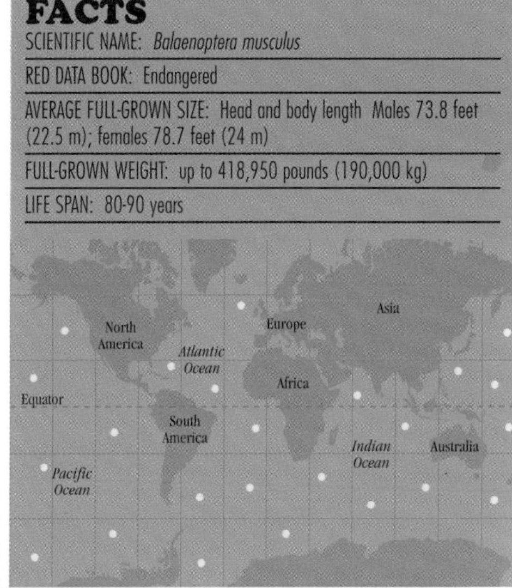

FACTS

SCIENTIFIC NAME: *Balaenoptera musculus*

RED DATA BOOK: Endangered

AVERAGE FULL-GROWN SIZE: Head and body length Males 73.8 feet (22.5 m); females 78.7 feet (24 m)

FULL-GROWN WEIGHT: up to 418,950 pounds (190,000 kg)

LIFE SPAN: 80-90 years

A close-up of a blue whale's massive tail.

FIN WHALE

This fin whale's eye is just visible near the corner of its enormous mouth.

The fin whale is the second largest species of whale after the blue whale. It lives in the open oceans, and rarely ventures into coastal waters. During winter, fin whales breed in temperate waters and then migrate to cold polar seas to feed during the summer months. Three main populations exist in the Southern Ocean, North Atlantic, and North Pacific.

Fin whales feed on krill and shoaling fishes such as herring, cod, mackerel, and capelin. The whales probably choose whatever food is available. Fin whales usually live in small groups of six or seven, but they have been sighted in groups of one hundred or more. Females give birth to a single calf every two or three years.

Fin whale hunting began in the North Atlantic just over one hundred years ago, when the harpoon cannon was invented. Huge numbers were killed in the Southern Ocean, especially after the blue whales became rare. Less than one quarter of the original population now live in the Southern Hemisphere, although there are many more in the Northern Hemisphere. Commercial hunting continued in Spanish and Icelandic seas until the 1980s.

The species is now protected from commercial whaling by the International Whaling Commission. Greenlanders are permitted to kill a few fin whales for their own needs. As with other whales, collisions with ships and drowning in fishing nets are significant threats to the survival of the species.

FACTS

SCIENTIFIC NAME: *Balaenoptera physalus*

RED DATA BOOK: Vulnerable

AVERAGE FULL-GROWN SIZE: Head and body length Males 62 feet (19 m); females 65.6 feet (20 m)

FULL-GROWN WEIGHT: 154,350 pounds (70,000 kg)

LIFE SPAN: 114 years

North America
Atlantic Ocean
Europe
Asia
Africa
Equator
South America
Indian Ocean
Australia
Pacific Ocean

HUMPBACK WHALE

The humpback whale is the best known of all the great whales. This is because it spends so much time near coasts and is so inquisitive that it often comes close to ships. Over eight thousand humpbacks have been identified individually by photographing the unique pattern on the tail flukes, which are raised in the air when the whale dives. The humpback was also the first great whale to be hunted.

Humpbacks live in all the world's oceans from the Arctic to the Antarctic. The populations in the Northern and Southern hemispheres, however, do not mix. The original population in the Southern Hemisphere was over 100,000 before hunting started. It is now about 3,000. There are even fewer humpbacks in the Northern Hemisphere. Like the other rorquals and the right whales, the humpback feeds by straining swarms of small animals from the water with its baleen or whalebone. Its main food is the shrimplike krill, but it also eats a variety of fish such as mackerel, capelin, and herring. Humpbacks breed in shallow water in the tropics. Their "songs" can be heard during the breeding season. Females bear a calf every one to three years. The single calf stays with its mother for six months or more.

Whale hunting was prohibited in the Southern Hemisphere in 1963 and in the North Pacific in 1966. Since then only a few whales have been killed each year by subsistence hunters in Greenland, the Cape Verde Islands, and elsewhere. Humpbacks are especially threatened by fishing nets because these whales swim near the shore. Special teams in Newfoundland rescue tangled humpbacks. The numbers of humpbacks are increasing in some places, and they have become tourist attractions where they regularly swim near the coast.

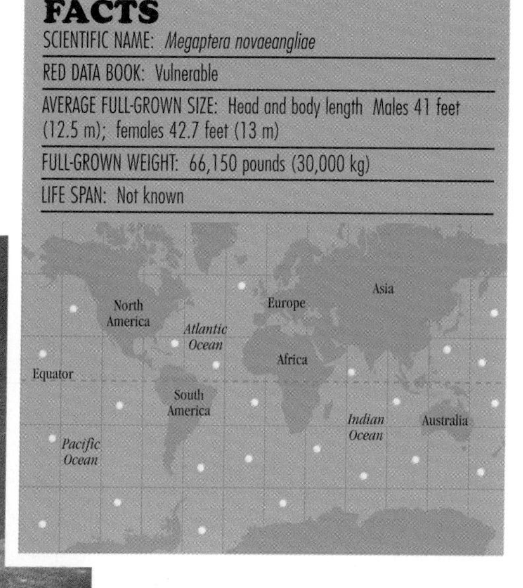

FACTS

SCIENTIFIC NAME: *Megaptera novaeangliae*

RED DATA BOOK: Vulnerable

AVERAGE FULL-GROWN SIZE: Head and body length Males 41 feet (12.5 m); females 42.7 feet (13 m)

FULL-GROWN WEIGHT: 66,150 pounds (30,000 kg)

LIFE SPAN: Not known

A humpback whale and calf swim together at the water's surface.

HECTOR'S DOLPHIN

This small dolphin is named after the New Zealand zoologist who discovered it in 1869. It is also called the white-headed dolphin, and it is found only in the seas around New Zealand. Half of all sightings are within 2,625 feet (800 m) of the shore, but the dolphins sometimes swim up rivers and go farther offshore during the winter.

Hector's dolphins live in small schools of two to eight animals, but they sometimes gather in larger groups. Not much is known about their habits. Individual dolphins can be recognized by their natural markings and are often seen in the same place for several years. They eat whatever fish they can find, especially small fish less than 14 inches (35 cm) long. They also eat squid and crustaceans. A single calf is born every two years.

The total population of Hector's dolphins is only 3,000 to 4,000, and it is declining. The dolphins used to be killed for fishing bait, but the recent decline is due to accidental tangling in fishing nets. The species is protected by New Zealand law, and all dolphins drowned in nets are supposed to be reported. Many are not reported, however, and experts believe that over one hundred Hector's dolphins may be killed in this manner each year off the coast of New Zealand.

A marine mammal sanctuary has been established at Banks Peninsula. All fishing nets are forbidden in the sanctuary for part of the year.

Hector's dolphins are usually found near the shoreline.

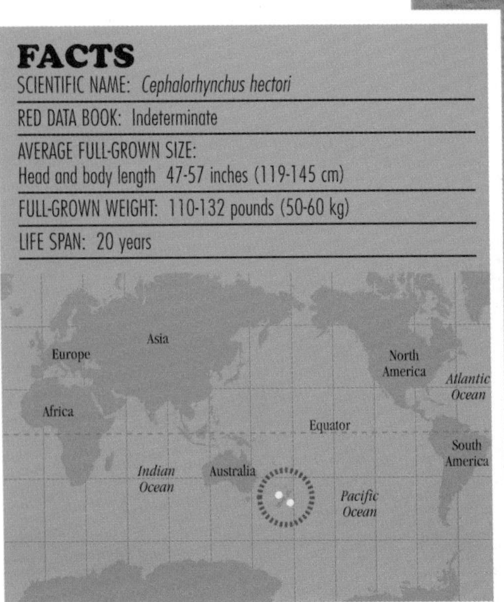

FACTS

SCIENTIFIC NAME: *Cephalorhynchus hectori*

RED DATA BOOK: Indeterminate

AVERAGE FULL-GROWN SIZE:
Head and body length 47-57 inches (119-145 cm)

FULL-GROWN WEIGHT: 110-132 pounds (50-60 kg)

LIFE SPAN: 20 years

NARWHAL

The male, and occasionally the female, narwhal has a single tusk that grows to 6.6 feet (2 m) in length, which was once believed to be the horn of the mythical unicorn. In fact, the tusk is an overgrown tooth.

Narwhals are found only in the Arctic Ocean north of the pack-ice boundary. They are social animals and are often seen in schools of either males or females and their calves. Narwhals usually stay in deep waters, but occasionally they come into shores and estuaries. They can break through new ice that is less than 2-2.4 inches (5-6 cm) thick, but they occasionally get trapped and drown when the sea freezes. Narwhals eat a variety of fish and some crustaceans and squid. The females bear a single calf every three years.

Inuit and other people who live around the Arctic Ocean hunt narwhals for their meat, blubber, and other products. The skin, called muktuk, is a delicacy that is rich in Vitamin C. Europeans have hunted narwhals ever since explorers sailed into the Arctic. The tusks have always been valuable because they are made of ivory and can be carved into luxury goods. People once believed the tusks could protect against poisoning. For this reason, thousands of narwhals were killed each year.

The world population of narwhals is not known, but there were 34,000 in Baffin Bay in 1994. Although the narwhal is protected, about 1,000 are killed every year. Local people can still hunt narwhals, but all kills must be recorded, and tusks can be sold only if they are "tagged" for identification. Scientists fear that too many narwhals are still being killed, and that the population will continue to decline.

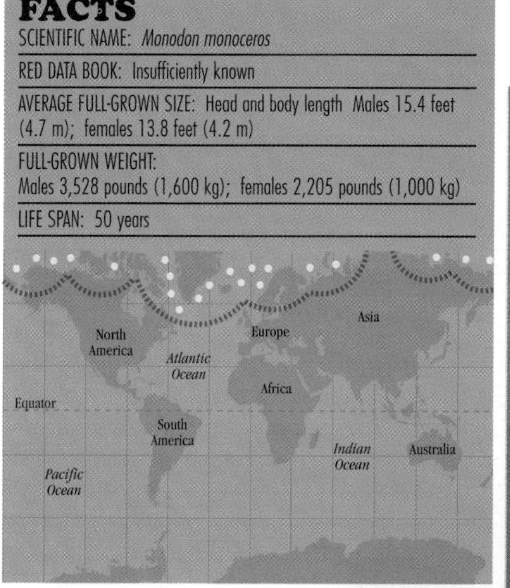

FACTS

SCIENTIFIC NAME: *Monodon monoceros*

RED DATA BOOK: Insufficiently known

AVERAGE FULL-GROWN SIZE: Head and body length Males 15.4 feet (4.7 m); females 13.8 feet (4.2 m)

FULL-GROWN WEIGHT: Males 3,528 pounds (1,600 kg); females 2,205 pounds (1,000 kg)

LIFE SPAN: 50 years

This aerial photograph shows a group of narwhals swimming in the Canadian Arctic.

VAQUITA PORPOISE

This rare photograph shows the dorsal fins of a female vaquita and her calf.

The vaquita is a porpoise that was only discovered as a new species in 1958. It is the most endangered of all the cetaceans (whales, dolphins, and porpoises) and one of the most endangered of all mammals. It is found only in the northern part of the Gulf of California, in Mexico. Only a few hundred of these porpoises remain, and the danger of entanglement in fishing nets is a constant threat to their survival.

The waters at the shallow northern end of the Gulf of California are teeming with food. The vaquita can exploit the abundance of food in this area through its ability to swim in lagoons so shallow that its back sticks out of the water. Vaquitas live in small groups, but scientists know very little about their habits.

The main threat to the vaquita has been the totoaba fishing industry. The totoaba is a fish that is also endangered. At one time, thirty to forty vaquitas were drowned every year in totoaba nets. Fishing for this species is now illegal, but poachers still set nets. Vaquitas also drown in nets set for other fish.

The rich habitat of the Gulf of California is threatened by the damming of the Colorado River, which prevents nutrient-rich water from flowing into the Gulf. The northern Gulf of California has been declared a Biosphere Reserve by the Mexican Government, and the navy enforces fishing regulations. Teaching local fishermen about the value and interest of the vaquita will be important in helping the porpoises survive.

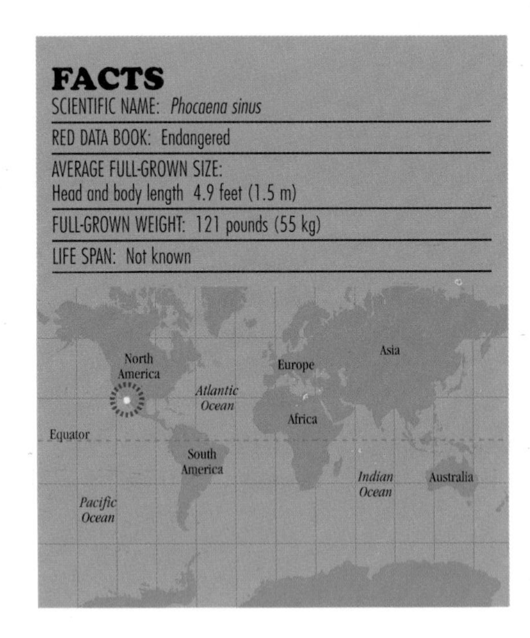

FACTS

SCIENTIFIC NAME: *Phocaena sinus*

RED DATA BOOK: Endangered

AVERAGE FULL-GROWN SIZE:
Head and body length 4.9 feet (1.5 m)

FULL-GROWN WEIGHT: 121 pounds (55 kg)

LIFE SPAN: Not known

BAIRD'S BEAKED WHALE

FACTS

SCIENTIFIC NAME: *Berardius bairdii*

RED DATA BOOK: Insufficiently known

AVERAGE FULL-GROWN SIZE:
Head and body length 33-39 feet (10-12 m)

FULL-GROWN WEIGHT: 17,640-22,050 pounds (8,000-10,000 kg)

LIFE SPAN: 80 years

So little is known about Baird's beaked whale that experts do not know whether the population is declining.

Baird's beaked whale is the largest of the whale family known as beaked and bottle-nosed whales, and it is named after an American naturalist. Its range covers the North Pacific, from the Pribilof Islands, Kamchatka, and Alaska southward to Japan and Baja California in Mexico. It prefers water more than 3,280 feet (1,000 m) deep and is not often seen near coasts. However, whole schools sometimes become stranded on beaches, although scientists do not know why this occurs.

Although humans have hunted Baird's beaked whale for many years, very little is known about its habits. The whales usually live in schools of three to twenty, but solitary animals can sometimes be seen. Their diet includes fish and squid that are often caught in deep water. Baird's beaked whale can dive regularly to 3,280 feet (1,000 m). Dives can be as deep as 7,874 feet (2,400 m) and last for over one hour.

At one time, Baird's beaked whales were hunted off British Columbia and California, and they are still hunted off Japan. Until 1971, over 200 were killed each year. Since then, smaller numbers have been killed, and the Japanese government sets a quota. Forty Baird's beaked whales are caught by whalers per year. Some beaked whales are also drowned in fishing nets. Because the size of the Baird's beaked whale population is not known (there are probably between 4,000 to 6,000 in the western North Pacific), it is impossible to determine if the population is declining.

DUGONG

The dugong is one of the sirenians, or sea cows. It is completely aquatic and cannot come ashore. Like the whale, the dugong has no hair, and its tail is flattened into a broad fluke for swimming. Dugongs are found in warm, sheltered water around the coasts of the Indian and western Pacific oceans. Their range is from Mozambique and Madagascar through the Red Sea and Persian Gulf around India to the islands of southeastern Asia as far north as Taiwan and the Ryukyu Islands of Japan and south around the coast of Australia.

The range of the dugong is restricted by the presence of the sea grasses that are almost its only food. This means that dugongs are only found where the sea is shallow and filled with fine silt for the grasses to grow in. When the sea grass is scarce, dugongs will sometimes eat seaweeds. Dugongs live in herds of up to several hundred, although some can be solitary. Their habits are difficult to study because they are active mainly at night and spend most of their time underwater. They breed very slowly, so they are vulnerable to extinction.

Dugongs have always been hunted for their meat, skin, and blubber by local people. Other parts of the body are also utilized. In Indonesia, dugongs were once hunted because their tears were used as good-luck charms. This type of hunting did not affect the numbers of dugongs until the development of modern firearms and motorboats. Many dugongs now drown in fishing nets as well as in the nets used to keep sharks away from bathing beaches. Polluted coastal water is also a growing problem.

Dugongs have disappeared from many parts of their range. Commercial hunting and trade are illegal, but many dugongs are killed anyway. Some reserves help protect the dugongs, such as the Paradise Islands National Park in Mozambique, and the Great Barrier Reef Marine National Park in Australia. More marine reserves are needed, however, to preserve dugongs in their favorite haunts.

FACTS

SCIENTIFIC NAME: *Dugong dugon*

RED DATA BOOK: Vulnerable

AVERAGE FULL-GROWN SIZE:
Head and body length 94.6-106 inches (240-270 cm)

FULL-GROWN WEIGHT: 507-794 pounds (230-360 kg)

LIFE SPAN: 70 years

North America
Atlantic Ocean
Europe
Asia
Africa
Equator
South America
Indian Ocean
Australia
Pacific Ocean

The dugong feeds almost entirely on sea grasses growing in shallow waters.

CARIBBEAN MANATEE

The Caribbean manatee and its relatives, African manatees and South American manatees, are all species of sirenians, or sea cows. When Christopher Columbus reached the West Indies, he thought manatees were mermaids. Caribbean manatees range from Virginia in the United States, to Brazil, but they now live only in scattered areas. Less than 2,000 remain in Florida.

Caribbean manatees live in warm, shallow waters. They are not completely marine and often live in estuaries and rivers. They eat a variety of vegetation, and they reach up to pull leaves from trees that hang over the water. At one time, manatees in Guyana helped keep irrigation channels clear of vegetation. Manatees are difficult to study because they remain under water most of the time. They often gather in groups, especially to feed, and move into warmer water in the winter. The females produce only one calf every three years.

Manatees are hunted for their meat, skin, and bones. They were once used as a cheap source of meat for workers on sugar plantations and, as a result, manatees have disappeared from some places.

Today, disturbance has become a serious problem for the manatees. New housing and industrial developments destroy the manatees' habitat with loud noise and pollution. The quiet coasts and inland waters, such as the Florida Everglades, are now filled with boats. Many manatees are badly cut by the propellers of powerboats.

Conservation of the Caribbean manatee depends on educating people, especially the drivers of powerboats, to be aware of manatees. Reserves to protect the manatees are also badly needed.

FACTS

SCIENTIFIC NAME: *Trichechus manatus*

RED DATA BOOK: Vulnerable

AVERAGE FULL-GROWN SIZE:
Head and body length 118-157.6 inches (300-400 cm)

FULL-GROWN WEIGHT: 1,102.5 pounds (500 kg)

LIFE SPAN: 50 years

North America
Atlantic Ocean
Europe
Asia
Africa
Equator
South America
Pacific Ocean
Indian Ocean
Australia

Manatees often live in estuaries and rivers as well as in the sea.

YELLOW-EYED PENGUIN

The yellow-eyed penguin nests around the southeastern coast of South Island in New Zealand, and on a few islands farther south: Stewart, Auckland, and Campbell islands. It docs not range far from these places. Outside the breeding season, penguins gather on beaches in between their daily trips to the sea in search of food.

Yellow-eyed penguins feed mainly on fish, but they also eat squid and some crustaceans. They nest singly or in small colonies in the dense forest vegetation or on grassy cliff tops. Females lay two eggs.

There are between 1,400 to 1,800 pairs of breeding yellow-eyed penguins. The species has been studied for many years, and its numbers change from year to year, but the reasons are not always clear. Between 1986 and 1987, the pairs on South Island dropped by 65 percent, probably because of a food shortage. One problem is that sheep grazing creates grassland where rabbits thrive. The abundance of rabbits attracts predators that also kill penguin chicks. Sometimes predation is as high as 90 percent on penguin chicks that nest in farmed areas. The penguins are also sensitive to disturbance by humans, and they will not come ashore if there are people on a beach. Yellow-eyed penguins are being helped by releasing captive-reared chicks in safe, predator-free areas, where they will breed as adults.

Yellow-eyed penguins are shy birds that nest in dense forest vegetation or on grassy cliff tops.

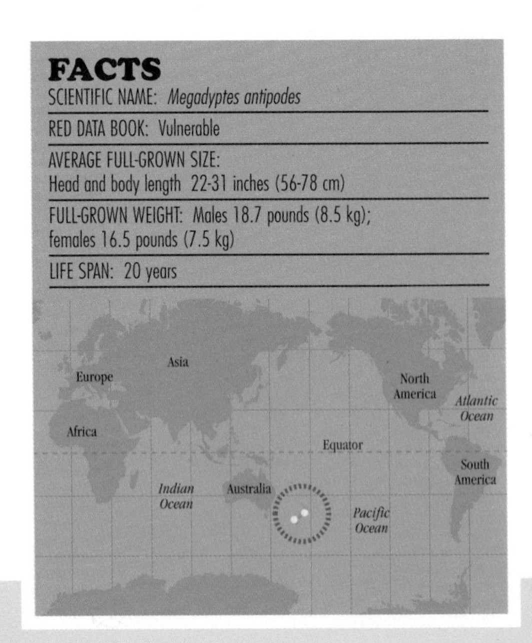

FACTS

SCIENTIFIC NAME: *Megadyptes antipodes*

RED DATA BOOK: Vulnerable

AVERAGE FULL-GROWN SIZE:
Head and body length 22-31 inches (56-78 cm)

FULL-GROWN WEIGHT: Males 18.7 pounds (8.5 kg); females 16.5 pounds (7.5 kg)

LIFE SPAN: 20 years

SHORT-TAILED ALBATROSS

At one time, the short-tailed albatross, also called Steller's albatross, flew over the sea almost everywhere in the North Pacific. Now it is rarely seen far from Japan. Huge nesting colonies once existed on many of the Izu-Bonin and Ryukyu islands of Japan, on islands off Taiwan, and probably several other islands off China's mainland. But by 1950, only 150 birds of this once abundant species remained alive.

Short-tailed albatrosses are long lived but slow breeding. They nest in tightly packed colonies and lay only a single egg each year. When not on their nests, they fly out to sea to feed on fish, squid, and crustaceans that they catch at the surface.

The dramatic decline of the short-tailed albatross was caused by the enormous market for its feathers, which were used in Japan and China for filling quilted clothing and bedding. Five million albatrosses were killed between 1872 and 1889. The species survived only on Tori-shima (which means "Bird Island"), one of the Izu Islands. Although the killing of albatrosses was banned as early as 1909, the collection of feathers stopped only in 1939 with the outbreak of World War II.

In 1950, short-tailed albatrosses were discovered breeding on Tori-shima again. With careful protection, numbers rose to 26 pairs in 1965. The population is now increasing at about 7 percent per year. In 1991, 500 short-tailed albatrosses lived on Tori-shima, 75 on Minami-kojima in the Senkaku Islands of Japan, and a pair on Midway Island in Hawaii.

The Japanese government has declared the short-tailed albatross a Special National Monument and has turned down a proposal to mine sulfur on Tori-shima. Albatrosses still face danger from drowning in fishing nets, from being hooked on fishing lines, and perhaps from predation by rats. Also, Tori-shima is a volcano, and an eruption could easily wipe out the main population.

A flock of short-tailed albatrosses feeds in the seas off Japan.

FACTS

SCIENTIFIC NAME: *Diomedea albatrus*

RED DATA BOOK: Endangered

AVERAGE FULL-GROWN SIZE:
Total length 33-37 inches (84-94 cm)

FULL-GROWN WEIGHT: 6.6-8.8 pounds (3-4 kg)

LIFE SPAN: Not known

AMSTERDAM ISLAND ALBATROSS

The Amsterdam Island albatross was discovered only in 1977. Some French ornithologists were studying wandering albatrosses on Amsterdam Island in the middle of the Indian Ocean. They realized that a small number of birds that they first thought were young wandering albatrosses had different nesting habits. They turned out to be a new species. Experts now know that there are only 70 Amsterdam albatrosses in the world. They all live on Amsterdam Island, and only ten pairs nest each year.

Amsterdam albatrosses nest only on the top of Amsterdam Island, at a height of 1,640-1,970 feet (500-600 m) above sea level. Evidence suggests that a larger population once existed, because the remains of the albatrosses have been found on other parts of the island. Like wandering albatrosses, Amsterdam albatrosses nest only every other year. They lay one egg, and the chick does not leave the nest until the next summer. It does not become mature for nine years. After nesting, the albatrosses fly huge distances over the ocean, sometimes reaching as far away as Australia.

Sealers, whalers, and fishermen used to land on Amsterdam Island, and they probably killed albatrosses for food. Cattle that roamed the island presented another threat. As the cattle destroyed the vegetation on the lower parts of the island, they moved uphill toward the albatrosses. A fence now keeps them out. Cats on the island may also kill some albatross chicks. In addition, fires started by humans have devastated parts of Amsterdam Island. If fires reach the top of the island, albatross nests will certainly be destroyed.

The safety of the Amsterdam Island albatross depends upon careful management of the island. Destruction of its nesting habitat and unchecked predators could easily wipe out the small population. If more areas of suitable habitat can be protected, the population may expand. This would, however, take a long time because Amsterdam albatrosses breed so slowly.

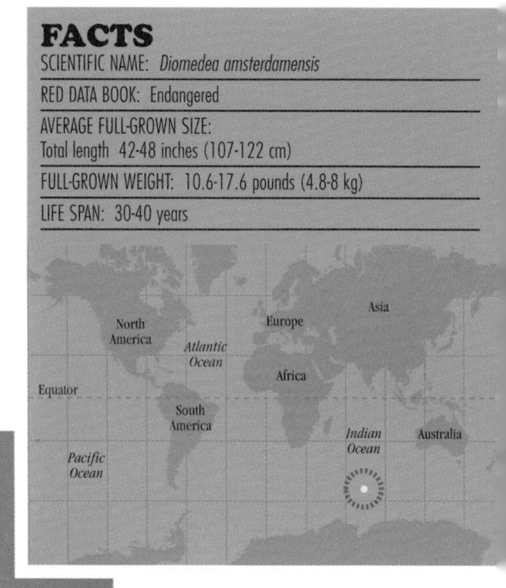

FACTS

SCIENTIFIC NAME: *Diomedea amsterdamensis*

RED DATA BOOK: Endangered

AVERAGE FULL-GROWN SIZE:
Total length 42-48 inches (107-122 cm)

FULL-GROWN WEIGHT: 10.6-17.6 pounds (4.8-8 kg)

LIFE SPAN: 30-40 years

The Amsterdam Island albatross is the rarest of the albatross family.

BERMUDA PETREL

No mammals lived on the islands of Bermuda when Europeans arrived, but there were large colonies of seabirds, including the Bermuda petrel. The birds were caught for food and killed by introduced pigs, cats, rats, and dogs. In 1619, the governor of Bermuda issued a proclamation to protect the cahow, as the Bermuda petrel was called. Two years later, the bird was believed extinct, and it was not seen again for three hundred years until a sighting in 1906. No nest was found until 1951; the population then numbered only eighteen pairs.

Bermuda petrels spend most of their time at sea, where they feed on small squid, fish, and crustaceans. Originally they nested in burrows in the soil, but these were very vulnerable to the introduced predators. The petrels now survive only on small islands that have no soil, only rock. They nest in cliff crevices or under boulders. Females lay a single egg.

The small population of Bermuda petrels currently faces other problems, such as increased development of Bermuda. Airports and other installations have caused nest sites to be abandoned. Hurricanes have taken their toll, and breeding has been harmed by pesticides.

At one time, up to 60 percent of Bermuda petrel chicks were dying as a result of competition with the white-tailed tropic bird that nests in the same crevices. Artificial entrances were made so the larger white-tailed tropic birds could not get in, and the breeding success of the Bermuda petrel increased dramatically. Artificial burrows have also been made on the cliff tops where the Bermuda petrels used to live, so the population can move away from the cliffs where the tropic birds nest.

In 1994, there were forty-five nesting pairs of Bermuda petrels. Scientists hope the birds will soon recolonize other islands.

FACTS

SCIENTIFIC NAME: *Pterodroma cahow*

RED DATA BOOK: Endangered

AVERAGE FULL-GROWN SIZE:
Head and body length: 13.8-15.8 inches (35-40 cm)

FULL-GROWN WEIGHT: Not known

LIFE SPAN: Not known

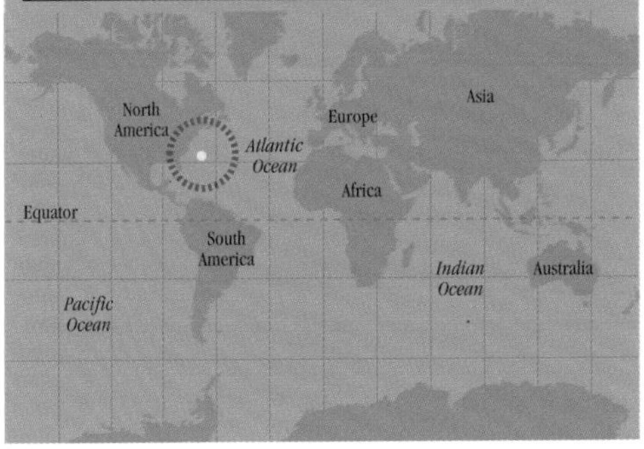

A sketch of the Bermuda petrel, thought to be extinct by 1621. The species was rediscovered in the twentieth century, and in 1994 there were at least forty-five breeding pairs in the wild.

MADEIRA PETREL

The Madeira petrel is one of the world's most endangered seabirds.

The Madeira petrel was discovered in 1903. It nests only in a small area of the central mountains of Portugal's Madeira Islands in the North Atlantic, although fossils show that the species was once more widespread. It is one of the most endangered seabirds in the world. There are now probably only twenty to thirty breeding pairs in existence.

Scientists do not know where Madeira petrels go when they are at sea. They are seen in Madeiran waters in the breeding season, but then they disappear. They feed on small fish, squid, and crustaceans. They make nest burrows in soil on cliff ledges at altitudes of 5,250 feet (1,600 m) in the mountains.

Sheep and goat grazing has destroyed large areas of Madeira petrel nesting habitat, and the species now breeds only in places too inaccessible for the grazers. At one time, shepherds ate the chicks, but this rarely happens now because the remaining burrows are too difficult to reach. Collectors who steal adults and eggs as specimens are more of a problem.

The main threat to the Madeira petrel comes from introduced cats and rats. They kill adults, eggs, and chicks. In 1991, cats killed ten adult birds.

At the moment, the nesting area is not legally protected, so it is impossible to fence off the sheep and goats. The Freira Conservation Project (Freira is a local name for the Madeira petrel), however, helps protect the species by setting out poison and traps for rats and cats on the nesting ledges. This helps the birds breed successfully.

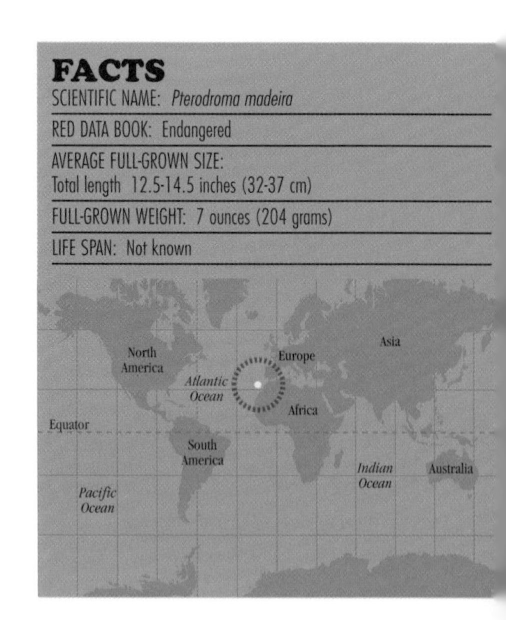

FACTS

SCIENTIFIC NAME: *Pterodroma madeira*

RED DATA BOOK: Endangered

AVERAGE FULL-GROWN SIZE:
Total length 12.5-14.5 inches (32-37 cm)

FULL-GROWN WEIGHT: 7 ounces (204 grams)

LIFE SPAN: Not known

ROUGH-FACED SHAG

A colony of rough-faced shags rests on a rock in the Marlborough Sounds in New Zealand.

The rough-faced shag nests only on four large rocks in the Marlborough Sounds, off Cook Strait between the North and South islands of New Zealand. At last count, the population totaled only 524 birds. Scientists do not know if the species range was ever more widespread, but the population was once larger.

Rough-faced shags remain near their nesting colonies all year. They feed in sheltered bays and coves within 15.5 miles (25 km) of the colonies and dive for bottom-living fish and crustaceans. The colonies are small, numbering two to eighty nests, and the nests are built of seaweed above the reach of the waves.

In the past, some rough-faced shags and their eggs were collected as museum specimens, and many more were hunted for their feathers. Recently, they have been shot by fishermen who thought the birds stole their fish. The species became protected in 1924, and few are killed today. The population is now steady, but nesting success is low. Out of ten eggs laid, only one chick survives to leave the nest. Visits by tourists and divers cause the loss of many eggs. The shags panic and desert their nests when boats come too close. This allows gulls to steal their eggs. An agreement now exists for boat owners to stay about 330 feet (100 m) away from the colonies, but this may not be enough to protect such a shy bird.

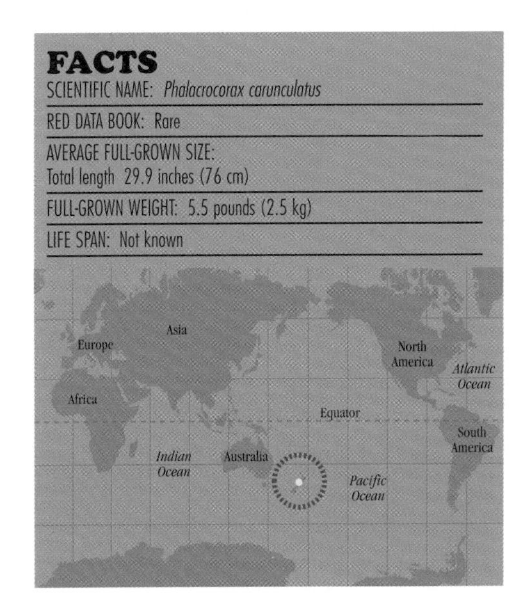

FACTS

SCIENTIFIC NAME: *Phalacrocorax carunculatus*

RED DATA BOOK: Rare

AVERAGE FULL-GROWN SIZE:
Total length 29.9 inches (76 cm)

FULL-GROWN WEIGHT: 5.5 pounds (2.5 kg)

LIFE SPAN: Not known

FLIGHTLESS CORMORANT

Flightless cormorants live only on two of the Galápagos Islands; the current population is about one thousand birds.

The flightless cormorant of the Galápagos Islands is, like all flightless birds living on remote islands, vulnerable to introduced predators. It lives only on the islands of Fernandina and Isabela and stays near its colonies, so the populations on the two islands do not mix. The total population is around one thousand birds, so a single disaster could make the species endangered or even extinct. The species is affected by El Niño, and, in 1983, the resulting climatic change caused the flightless cormorant population to drop by one-half.

Flightless cormorants nest among rocks on sheltered beaches, cliffs, and lagoons. Each colony consists of a few pairs that build nests of seaweed just above the high-water mark. They lay two or three eggs. The adults do not swim more than .6 mile (1 km) from the colony to feed. They dive for fish and octopuses that live on the seabed.

The remote and inaccessible colonies of the flightless cormorant save it from disturbance from the growing number of tourists visiting the Galápagos Islands. Introduced predators have been less of a problem since dogs were removed from Isabela. Increasing dangers for the cormorants include pollution from oil spills by passing ships and drowning in fishing nets.

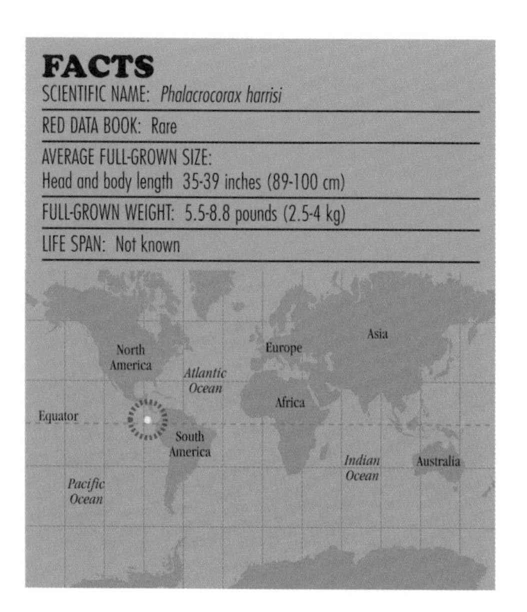

FACTS

SCIENTIFIC NAME: *Phalacrocorax harrisi*

RED DATA BOOK: Rare

AVERAGE FULL-GROWN SIZE:
Head and body length 35-39 inches (89-100 cm)

FULL-GROWN WEIGHT: 5.5-8.8 pounds (2.5-4 kg)

LIFE SPAN: Not known

CHRISTMAS ISLAND FRIGATEBIRD

The Christmas Island frigatebird nests only in a small area of forest on Christmas Island, south of Java. When not nesting, it stays at sea and sometimes goes as far as Australia or Hong Kong.

Frigatebirds are superb fliers and spend most of their lives on the wing over the sea searching for food. They eat flying fish and other marine animals they catch at the sea's surface. They also prey on eggs and chicks of other seabirds or steal food the adults bring back to their nests. Ten to twenty pairs of frigatebirds form a colony and nest together in tall trees. Each pair lays a single egg. It takes seventeen months to rear a chick, so breeding is very slow.

There are now fewer than 1,600 pairs of Christmas Island frigatebirds. The population used to be larger, and the birds once nested over a wider area of Christmas Island. They have suffered because the trees in their forest habitats have been cut down. Humans also hunted the birds for food, and phosphate mining was a problem because of the thick dust that covered the trees. Only three breeding colonies still exist. Deforestation and mining have stopped, and two of the colonies are in a national park, so the species may be safe. However, mining may possibly begin again. There is always a danger that a cyclone could destroy the nests and nesting trees. For such a small, slow-breeding population of birds, such an event could be very serious.

Christmas Island frigatebirds nest in tall trees within a small area of forested land.

FACTS

SCIENTIFIC NAME: *Fregata andrewsi*

RED DATA BOOK: Vulnerable

AVERAGE FULL-GROWN SIZE:
Head and body length 35-39 inches (89-100 cm)

FULL-GROWN WEIGHT: Males 3.1 pounds (1.4 kg); females 3.4 pounds (1.6 kg)

LIFE SPAN: Not known

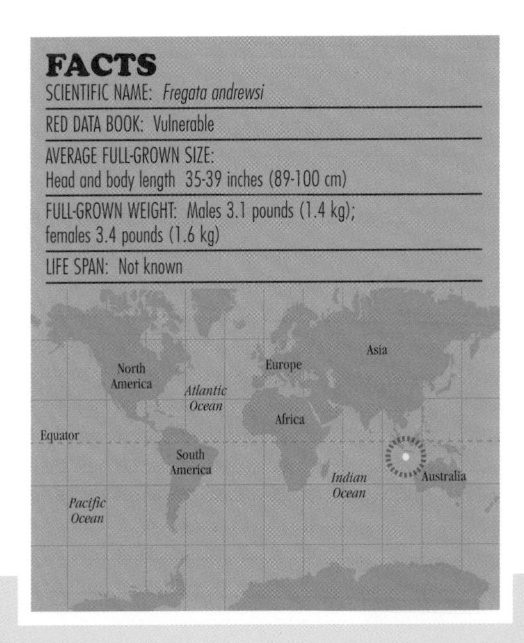

AUDOUIN'S GULL

A photograph of an Audouin's gull on an island off Turkey.

Audouin's gull lives around the Mediterranean Sea and is the world's rarest gull. Breeding colonies exist in most of the countries surrounding the Mediterranean, but the largest are in Spain. The gulls spend the winter along the coast of North Africa, from Libya westward to Morocco and down the Atlantic coast to Senegal and Gambia.

At one time, Audouin's gull was much rarer, but its numbers have risen in the last thirty years. In 1966, there were only 1,000 nesting pairs, but this had risen to over 9,000 by 1989. Nine out of ten pairs now nest in two Spanish colonies, the Ebro Delta and the Chafarinas Islands off Morocco. In 1992, the Spanish population was 12,631 pairs out of a total of under 14,000 pairs.

Audouin's gulls nest on small, rocky islands, although the Ebro Delta colony is on a sandy peninsula. They build their nests on stony slopes hidden among low vegetation. The birds lay a clutch of two or three eggs. Audouin's gulls catch small fish from the sea's surface, usually at night. They also follow fishing boats to feed on the garbage thrown overboard.

Egg collecting was the main reason for the Audouin's gull's rare status, but this has now almost stopped. The Chafarinas Islands became a reserve in 1982. A new threat comes from the yellow-legged gull, whose numbers are increasing rapidly. This gull competes for nesting places with Audouin's gull and also eats its eggs and chicks. Some of the yellow-legged gulls at the Chafarinas Islands are killed by humans each year, allowing Audouin's gulls to breed.

PAINTED TERRAPIN

The painted terrapin is also known as the Malaysian river turtle.

FACTS

SCIENTIFIC NAME: *Callagur borneoensis*

RED DATA BOOK: Endangered

AVERAGE FULL-GROWN SIZE:
Head and body length 19.5-27.5 inches (50-70 cm)

FULL-GROWN WEIGHT: 3.75 pounds (1.7 kg)

LIFE SPAN: Not known

River turtles are found throughout the warmer parts of the world. The painted terrapin lives in southeastern Asia, from southern Thailand in the north through Malaysia to Sumatra and Borneo. The painted terrapin makes its home in the tidal sections of rivers and mangrove swamps.

Painted terrapins spend much of their time resting with their heads out of the water or basking in the sunshine. Their main foods are fruit and leaves from plants growing over the water. Sometimes they feed on refuse, such as scraps of fruit thrown into the water from riverside villages. Females lay their eggs in sandbanks at low tide. Clutches contain up to twenty eggs.

Although they are still widely distributed, painted terrapins have disappeared from many rivers and mangrove swamps and are rare in the places where they survive. The species is almost extinct in Thailand, and only two or three rivers in Malaysia hold more than one hundred nesting females.

The most serious threat to the terrapins is the popularity of their eggs, which are worth more than four or five times the price of chicken eggs. The small number of eggs in a clutch and the ease of finding the nests make the painted terrapin easy to wipe out. In some places, the adult terrapins are caught as well. Athletes in Malaysia sometimes drink the terrapin's blood because they believe it will improve their performance. Another serious threat is the destruction of the nesting places by sand mining and construction work.

Painted terrapins have legal protection in some places, and egg collecting requires a license. In western Malaysia, egg collectors must give 70 percent of their eggs to the Fisheries Department so they can be hatched and the young terrapins released.

LOGGERHEAD TURTLE

The loggerhead turtle lives in tropical and subtropical seas around the world. At one time, it lived as far north as Virginia, in the United States, and it still lives as far south as the River Plate, in Argentina, and Natal, in South Africa. Occasionally, it wanders into temperate seas, and one was found at Murmansk, in northern Russia. The loggerhead is unusual among sea turtles because it swims up rivers.

Breeding beaches for the loggerhead are mostly in subtropical areas. Zoologists have recently discovered that the turtles hatched in Japan swim across the Pacific Ocean to feed at Baja California in Mexico. The journey takes two years. Eventually they swim back to Japan to breed. Loggerheads are carnivores and eat fish, shellfish, and jellyfish, as well as seaweed. Their jaws are strong enough to crush the shells of giant clams. They also eat Portuguese men o' war. Loggerheads nest on beaches guarded by rocks and reefs. Females dig a hole in the sand and deposit about 120 eggs.

Loggerheads are not caught for their meat, which has an unpleasant taste. Their eggs, however, have always been in demand. Loggerheads today are also threatened with noise disturbance when laying their eggs, because building developments serve to destroy breeding beaches. Both adult females and hatchlings are disoriented by bright lights when they head inland, and they are crushed while crossing roads. Fishing nets also entangle and drown many turtles. More than one thousand young Japanese loggerheads die in nets each year as they swim across the Pacific.

Conservation of the loggerhead turtle depends upon protected beaches, where they can breed in safety, and protection from fishing nets. In Oman, commercial hunting and collection of eggs is forbidden, and even local exploitation is controlled by law. People are encouraged to take eggs only from nests below the high-tide mark, where the eggs would be submerged.

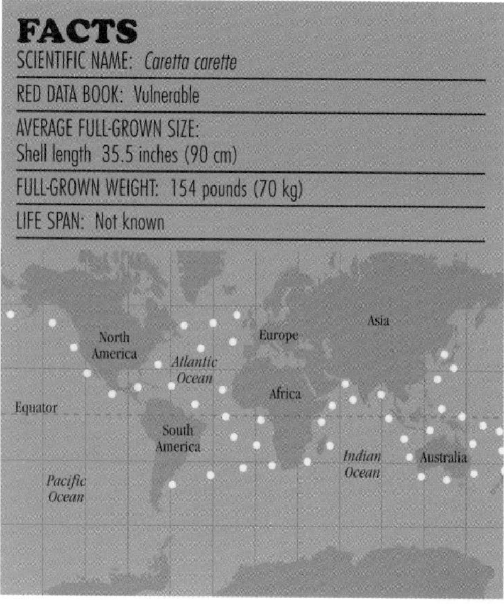
FACTS

SCIENTIFIC NAME: *Caretta carette*

RED DATA BOOK: Vulnerable

AVERAGE FULL-GROWN SIZE:
Shell length 35.5 inches (90 cm)

FULL-GROWN WEIGHT: 154 pounds (70 kg)

LIFE SPAN: Not known

A loggerhead turtle swims in the open sea at Langana Bay, Zákinthos, Greece.

GREEN TURTLE

The green turtle gets its name from its green fat. The species was once very abundant in all warm seas, but it has been overhunted and has disappeared or become rare in most places. Nesting beaches once existed around the world, but some, such as those in the Caribbean, have disappeared completely. Only about a dozen beaches exist today, with as many as 2,000 nesting female green turtles. Ten thousand turtles nest on Europa Island in the Mozambique Channel and tens of thousands on Raine Island in Australia.

Green turtles migrate between their feeding grounds, where they feed on sea grasses and seaweeds growing in shallow water, and their breeding beaches. The migration of green turtles from the coast of Brazil to tiny Ascension Island in the middle of the Atlantic Ocean, 1,242.8 miles (2,000 km) away, is one of the wonders of the natural world. Females lay three clutches of eggs per year, but they breed only once every three years, and they do not start laying until they are 15 to 50 years years old.

Turtle meat and eggs have always been important foods for people living near the turtles' feeding and nesting places. The turtles were also caught for their skins, shells, and oil. However, green turtles began to become rare when human populations in the tropics started to rise rapidly. Old customs that limited the harvest of turtles disappeared and were replaced by overexploitation. Many more turtle products were sold commercially, including preserved hatchlings sold as souvenirs to tourists. Fishing nets have also been responsible for drowning many green turtles.

Conservation will depend on protecting adult females, their eggs, and their hatchlings on the breeding beaches, as well as the turtles that swim offshore. Some breeding beaches are located in reserves, but protection laws are not always properly enforced. International trade in green turtles and their products is forbidden, but smuggling continues.

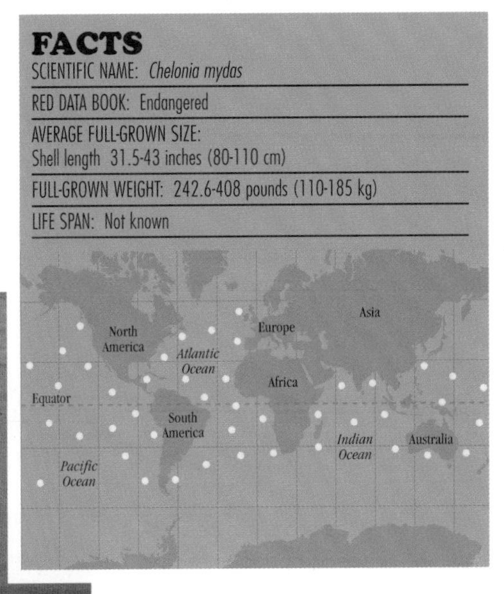

FACTS

SCIENTIFIC NAME: *Chelonia mydas*

RED DATA BOOK: Endangered

AVERAGE FULL-GROWN SIZE:
Shell length 31.5-43 inches (80-110 cm)

FULL-GROWN WEIGHT: 242.6-408 pounds (110-185 kg)

LIFE SPAN: Not known

A green turtle comes ashore at Sandpit Beach in Karachi, Pakistan.

HAWKSBILL TURTLE

This photograph of a hawksbill turtle emphasizes the size of the forelimbs, which are adapted for a lifetime of swimming.

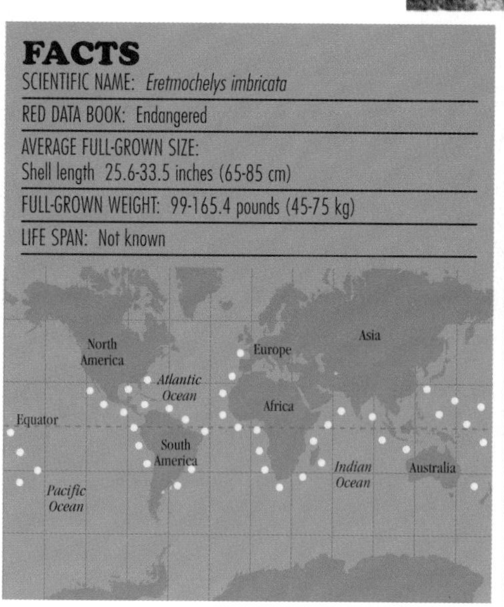

The hawksbill turtle is the source of tortoiseshell, which is used in many parts of the world for jewelry and other ornaments. Tortoiseshell is made from the plates of the turtle's shell, which are polished to show the attractive patterns. Hawksbill turtles live in all tropical seas, particularly in shallow waters. As with other marine turtles, they are not easy to study except when they come ashore to lay their eggs.

Experts are not certain if hawksbill turtles migrate long distances like some other marine turtles. It is possible they just keep swimming until they find a good place for feeding and then remain there. The turtles eat sponges and other forms of animal life growing on reefs. They nest on small sandy beaches, sometimes among other turtle species. Each female lays one to four clutches of 100 to 160 eggs per year. Each female may lay her eggs in only one year out of every three or four years.

Hawksbill turtles have become rare because of the tortoiseshell trade. The shells are now so valuable that people hunt the turtles even where they are rare. Most of the tortoiseshell goes to Japan. People also purchase polished shells and stuffed baby turtles, and they eat both the eggs and the adult turtles. Another problem centers around habitat loss where beaches are being destroyed or disturbed by tourism and industrial developments. Strict control of the tortoiseshell trade is necessary if the hawksbill turtle is to be saved. Setting up reserves around the reefs where the hawksbill turtles feed will also help stop the killing.

KEMP'S RIDLEY TURTLE

Kemp's ridley turtle is the most endangered of all sea turtles. It used to be common from New England down the coast of the United States and Mexico. This distribution is different from other sea turtles, most of which can be found all around the world.

Kemp's ridley turtles live in shallow inshore water. They feed mainly on bottom-living animals, such as shellfish and sea urchins, as well as fish and jellyfish. Crabs are a favorite food. Female Kemp's ridley turtles come ashore in swarms, called *arribadas* (Spanish for "arrivals"), to lay their eggs. A clutch averages 110 eggs. Each female lays eggs every year or two, instead of every three or more years as in other turtle species.

In the 1950s and 1960s, Kemp's ridley turtles became rare for several reasons. Humans and animals stole increasing numbers of eggs. Adult turtles, especially the females that came ashore to lay their eggs, were killed for food by humans. Also, large numbers were accidentally caught in fishing nets. No more than about 900 breeding females presently exist. Nearly all of these nest in one 12-mile (20-km) length of beach near Rancho Nuevo in northeastern Mexico. The main feeding grounds are on the coasts of Louisiana and Mexico.

Kemp's ridley turtle is protected by the Mexican government, and armed patrols guard the nesting beach. Another colony has been established near Corpus Christi, Texas, by rearing hatchlings from Rancho Nuevo and releasing them at their new home. Many turtles drown in nets because they feed in the same waters where shrimpers work. Therefore, fishing nets must be fitted with Turtle Excluder Devices (TEDs) that stop the turtles from drowning. The numbers of Kemp's ridley turtle are now increasing as a result of this precaution.

A Kemp's ridley turtle lays eggs on a beach in Mexico.

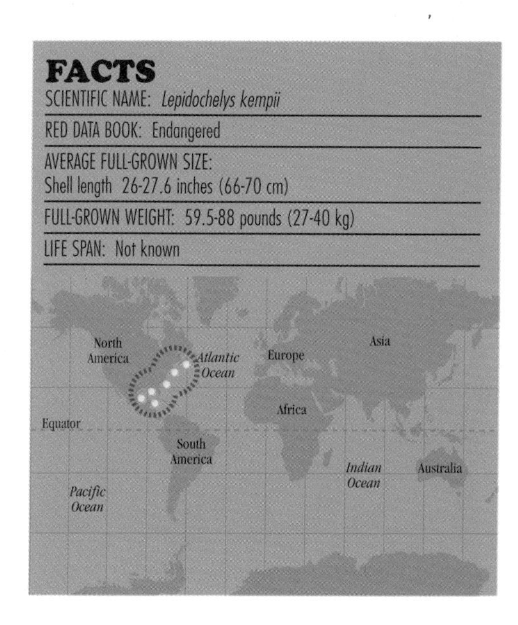

FACTS

SCIENTIFIC NAME: *Lepidochelys kempii*

RED DATA BOOK: Endangered

AVERAGE FULL-GROWN SIZE:
Shell length 26-27.6 inches (66-70 cm)

FULL-GROWN WEIGHT: 59.5-88 pounds (27-40 kg)

LIFE SPAN: Not known

OLIVE RIDLEY TURTLE

In contrast to Kemp's ridley turtle, the olive ridley turtle is probably the most abundant sea turtle. It is also the most hunted. The olive ridley turtle lives in shallow waters in tropical areas around the world, except for Hawaii and many other islands in the Pacific and the Caribbean.

Olive ridley turtles eat a variety of animals, including crabs, shrimp, jellyfish, and fish eggs. They may dive to over 328 feet (100 m) to find food on the seabed. Where they are still abundant, the females come ashore in masses called *arribadas*. Clutch size is about 110 eggs.

Although the olive ridley turtle remains widespread, the numbers of turtles at many breeding beaches have dwindled and almost disappeared in some places. Reasons include the hunting of adults, egg collecting, and accidental drowning in shrimp nets. In recent years, trade in turtle skins has developed, since the skin can be fashioned into a luxury leather. Although the species is often protected by law and its breeding beaches are in reserves, poaching continues to reduce the numbers of olive ridley turtles.

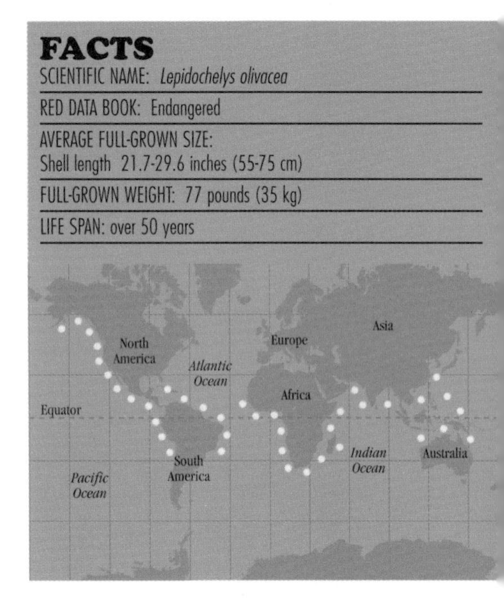

The olive ridley turtle is probably the most hunted sea turtle.

LEATHERBACK TURTLE

The leatherback is the largest sea turtle, and it gets its name from its tough, rubbery shell. Unlike other sea turtles, it lives in the open sea. The leatherback is a powerful swimmer and can be found far from shore. Although it nests on tropical beaches, it has a wide range, and many leatherbacks come into cool waters. They have been seen off Newfoundland and Norway in the north and New Zealand in the south.

Leatherbacks eat jellyfish and similar slow-swimming animals, along with the fish and shrimp that often live with them. Some leatherbacks have a regular annual migration into cool waters that have seasonal gatherings of jellyfish. Nesting beaches are on exposed coasts where the shore is steep and not protected by reefs. Females come to the waters off the nesting beaches every two to three years. They emerge to nest and lay eggs four to six times during that nesting season. The average clutch size is 85 eggs.

Egg collecting is a serious problem for leatherbacks. The adult turtles are not eaten often because their flesh is oily. However, exploitation of adults and eggs rises as human populations rise. A unique problem for leatherbacks is that they cannot easily distinguish between jellyfish and the increasing numbers of plastic bags floating in the sea. As many as half may have plastic bags in their stomachs, which could prove fatal.

The leatherback turtle is protected by law in most of its range, and many breeding beaches are in reserves. Egg collecting may still be permitted with a license. One Malaysian project specifies that all eggs must be collected on the same beach. Many of these eggs are then purchased from the collectors and allowed to hatch so the young turtles can be released in the sea.

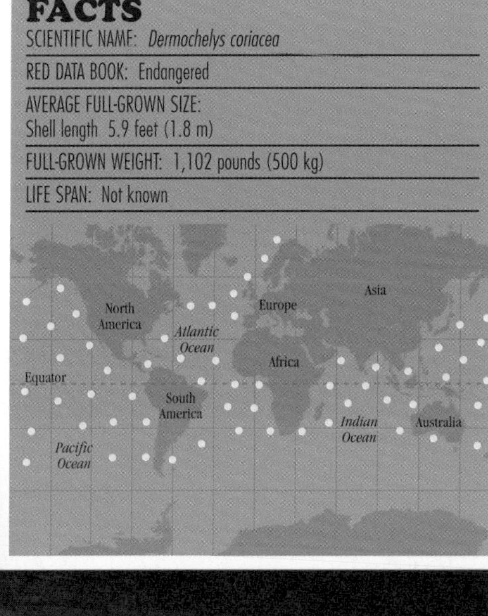

FACTS

SCIENTIFIC NAMF: *Dermochelys coriacea*

RED DATA BOOK: Endangered

AVERAGE FULL-GROWN SIZE:
Shell length 5.9 feet (1.8 m)

FULL-GROWN WEIGHT: 1,102 pounds (500 kg)

LIFE SPAN: Not known

A female leatherback turtle lays eggs on a beach in Trinidad.

ESTUARINE CROCODILE

The estuarine crocodile lives in the brackish waters of estuaries and mangrove swamps. It also inhabits rivers up to 93 miles (150 km) inland, freshwater pools, and swamps. The original range was from Cochin in western India through southeastern Asia to southern China and northern Australia. It now ranges from eastern India and Sri Lanka to the Solomon Islands, Vanuatu, and northern Australia, and the species is much rarer in all parts of this range. Occasionally, estuarine crocodiles are found in the Fiji Islands and Cocos-Keeling Islands.

The diet of an estuarine crocodile depends on its size. Very young crocodiles eat small fish, crustaceans, and insects. Larger juveniles eat larger fish and small snakes, mammals, and birds. Large adults can catch cattle and humans when they come to the water's edge.

Female estuarine crocodiles start to breed when they are ten years old. Their nests are mounds of plants and mud, often built on mats of floating vegetation. They lay about fifty eggs and stay near the nest to protect it. This makes the crocodiles vulnerable to human hunters.

Large populations of estuarine crocodiles now exist only in northern Australia and parts of New Guinea. Elsewhere they are rare, with only 170 to 330 in India. Habitat destruction along rivers and coasts has seriously reduced the range of the estuarine crocodile, and it has been hunted in large numbers for its skin. In the 1950s and 1960s, crocodile skin suddenly became very valuable, and hunters used high-power rifles and fast motorboats to increase their catch. Hundreds of thousands of estuarine crocodiles were killed each year. Hunting continues despite protection because a single crocodile skin may be worth half a year's normal wages.

Where protected, as in Australia's Kakadu National Park and India's Bhitarkanika Wildlife Sanctuary, the estuarine crocodile population is increasing. Rearing eggs in hatcheries and releasing the young crocodiles into the wild has also helped.

The skins of estuarine crocodiles have been fashionable since the 1950s.

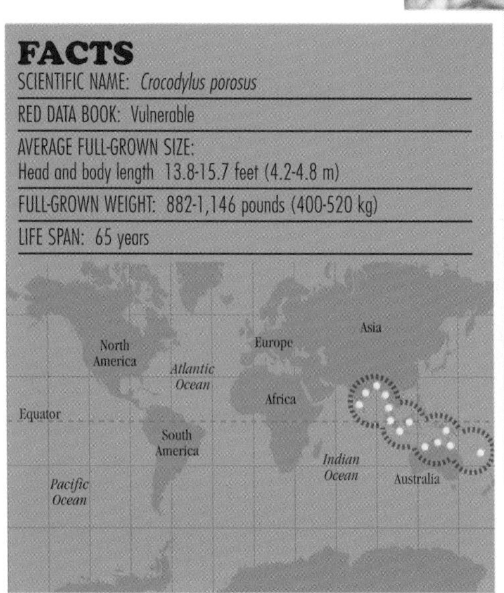

FACTS

SCIENTIFIC NAME: *Crocodylus porosus*

RED DATA BOOK: Vulnerable

AVERAGE FULL-GROWN SIZE:
Head and body length 13.8-15.7 feet (4.2-4.8 m)

FULL-GROWN WEIGHT: 882-1,146 pounds (400-520 kg)

LIFE SPAN: 65 years

GALÁPAGOS MARINE IGUANA

The Galápagos marine iguana only goes out to sea to feed.

FACTS

SCIENTIFIC NAME: *Amblyrhynchus cristatus*

RED DATA BOOK: Rare

AVERAGE FULL-GROWN SIZE: Head and body length Males 29.6 inches (75 cm); females 23.6 inches (60 cm)

FULL-GROWN WEIGHT: 1.1- 3.3 pounds (.5-1.5 kg)

LIFE SPAN: Not known

The Galápagos marine iguana is one of the large family of iguana lizards. It is now restricted to just a few islands in the Galápagos group. Scientists are not certain of the exact population, but 5,000 iguanas live on a .6 mile (1 km) stretch of coastline on the island of Santa Fé. Although abundant in some places, marine iguanas are considered at risk because of the species' small range. A single disaster could be serious.

Most of a marine iguana's life is spent on land, and it goes to sea only to feed. The sea around the Galápagos Islands is cold, and the iguanas do not feed until they have warmed up in the sun. They eat red and green seaweeds. The females and young eat while standing on rocks exposed at low tide, while the males dive to feed below the low tide mark. The females nest in colonies, and each one lays only two to four eggs per year.

Like many animals living on islands, marine iguanas have suffered from introduced predators such as cats, pigs, and rats. These are a serious threat because the iguanas lay so few eggs. Where the predators are kept under control or the iguanas are given safe nesting places, their numbers have risen. Other threats include oil spills that kill the seaweeds. About every twelve years, animal life on the Galápagos Islands suffers from the effects of El Niño, a warm ocean current that causes the sea temperature around the islands to rise. In 1982, the high temperature allowed a new species of seaweed to invade the Galápagos Islands, replacing the native species. The marine iguanas could not digest the new seaweed, and many died of starvation.

WHALE SHARK

The whale shark is the largest fish in the world. It lives in the warm, tropical waters of the Atlantic, Pacific, and Indian oceans. Whale sharks migrate and their numbers are small, so it is difficult to know how many exist and whether they are becoming rare. In some places, they are not sighted as often as they used to be.

Despite their large size, whale sharks are harmless to humans. The greatest threat to humans from whale sharks is through boat collisions. The sharks feed on small floating animals, mainly tiny crustaceans and fish larvae, but occasionally fish as large as sardines and anchovies. They feed by swimming slowly at the surface with their mouths open and straining swarms of these animals out of the water with their gills.

Whale sharks are caught all around the world, but usually in small numbers by local fishermen. The Taiwanese eat the most whale sharks, but the flesh is not very tasty. Recently, there has been a growth in the sale of shark fin soup, which has increased the number of whale sharks killed. Whale sharks are very shy and easily disturbed by divers who like to ride on their backs. This can upset the sharks' feeding because it drives them away from the area.

Fishing for whale sharks has been banned in the Maldives because the species population has declined. In Australia, whale sharks are protected in the Ningaloo Reef marine park. Many people come to see the whale sharks, but the numbers of boats and swimmers allowed around a shark are limited.

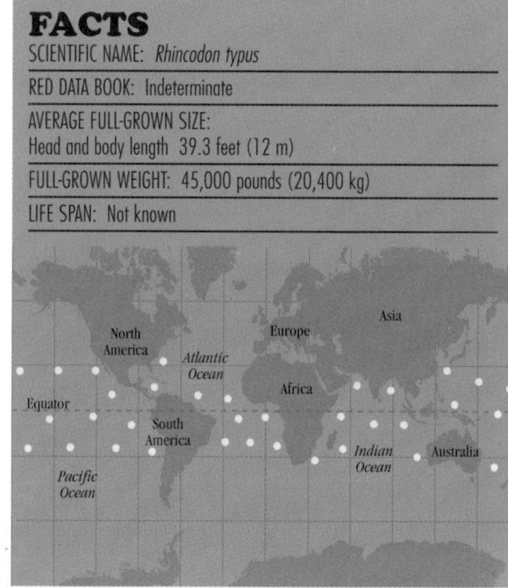

FACTS

SCIENTIFIC NAME: *Rhincodon typus*

RED DATA BOOK: Indeterminate

AVERAGE FULL-GROWN SIZE:
Head and body length 39.3 feet (12 m)

FULL-GROWN WEIGHT: 45,000 pounds (20,400 kg)

LIFE SPAN: Not known

Whale sharks are rare and difficult to monitor.

GREAT WHITE SHARK

Great white sharks are superbly adapted predators, although they rarely make unprovoked attacks on humans.

The great white shark is infamous as the man-eating shark in the movie *Jaws*. Although it is a super-predator, it is not as dangerous as most people think. Swimmers are far more likely to die through drowning or in diving accidents than from an attack by a great white shark. Great white sharks live in warm seas around the world. They can be spotted around coastlines but rarely far into the ocean.

Great white sharks are solitary, except when several gather to feed. Because it is at the top of the marine food chain, the species is not abundant. The sharks' main prey includes a variety of fish, such as pilchards, mackerels, salmon, and other sharks. The great white sharks also eat seals, porpoises, sea otters, and turtles, as well as shellfish, crabs, and squid.

Great white sharks probably live in separate territories, because individuals become familiar after a while. Scientists know very little about their habits. The females probably give birth to live young, and there is a record of one litter of nine young. Breeding is probably very slow.

It has been impossible to record any decrease in the population of the great white shark, but the low numbers and slow breeding rate mean it could easily become endangered. A few great white sharks are caught by humans for food. The skin is used for leather, and the liver yields an oil rich in vitamins. Because of the great white shark's evil reputation, its teeth and jaws can be sold for a high price, and it is the ultimate catch for shark fishers. The great white shark is protected in California and South Africa. Despite the shark's frightening reputation, its protection has received public support.

BASKING SHARK

Basking sharks feed by swimming with their mouths wide open to trap small fish and crustaceans in their gills.

FACTS

SCIENTIFIC NAME: *Cetorhinus maximus*

RED DATA BOOK: Insufficiently known

AVERAGE FULL-GROWN SIZE:
Head and body length 32.8 feet (10 m)

FULL-GROWN WEIGHT: Not known

LIFE SPAN: Not known

The basking shark is in the same family of sharks as the great white shark, but it is completely harmless. It gets its name from its habit of "basking" at the surface of the sea in fine weather. Basking sharks live in temperate and cold waters, from northern Norway and British Columbia to Florida and California in the Northern Hemisphere. In the Southern Hemisphere, they live around Australia, South America, and South Africa.

Basking sharks disappear in winter, probably because they migrate into deep water and stop feeding. Like the whale shark, the basking shark feeds on small animals. It swims with its mouth wide open and traps crustaceans and small fish on its gills. Sometimes large groups of basking sharks are sighted. There may be as many as one hundred sharks gathered where food is plentiful. Scientists know very little about their breeding habits except that the young are born live.

There have been fishing industries for basking sharks for centuries. The sharks have large livers that are rich in oil, and their skins can be made into leather. The flesh is made into fish meal. Basking sharks are also killed by fishermen when the sharks destroy fishing nets. The number of sharks killed is not very high, and no one knows whether the population has declined. Because there is no evidence that the species is threatened, it is difficult to arrange protection, although in parts of the British Isles, basking sharks are protected because they are a tourist attraction.

COELACANTH

Scientists once believed fossil records indicated that coelacanths became extinct sixty million years ago. However, a living coelacanth was caught by a fisherman near the Chalumna River in South Africa in 1938. It looked exactly like the fossils and is often described as a "living fossil." No further coelacanths have been found in South African waters, but some have since been found near the Comoros Islands, 995 miles (1,600 km) away. The coelacanths live in deep water, at about 655 feet (200 m), which makes them difficult to study, but miniature submarines now make this possible.

Coelacanths live in caves on the seabed. They feed on small fish and cuttlefish. Females give birth to about five live young each year. Because they produce so few young, the population increases slowly, and the species is at risk.

Since 1987, when scientists started to study live coelacanths, there has been a serious decline in their numbers. The scientists counted only one-third as many coelacanths in one area in 1994 as they had seen in 1989. In that year, motorboats replaced the canoes used by local fishermen. The fishermen were able to catch many more fish and make more money. Then the motors broke down, and the fishermen had to fish where the coelacanths live to maintain their catch. Unfortunately, they caught large numbers of coelacanths. The plan now is to improve the fishing near shore, where there are no coelacanths.

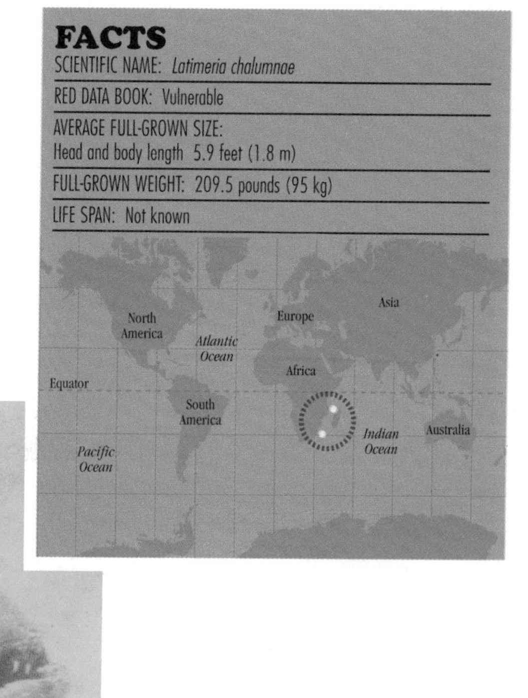

FACTS

SCIENTIFIC NAME: *Latimeria chalumnae*

RED DATA BOOK: Vulnerable

AVERAGE FULL-GROWN SIZE:
Head and body length 5.9 feet (1.8 m)

FULL-GROWN WEIGHT: 209.5 pounds (95 kg)

LIFE SPAN: Not known

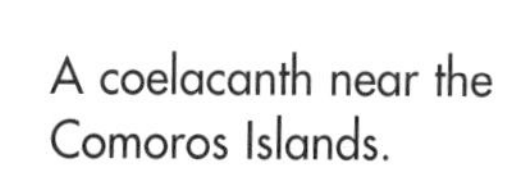

A coelacanth near the Comoros Islands.

COMMON STURGEON

Caviar is the eggs, or roe, of the common sturgeon. This delicacy is very valuable to humans, which has caused the sturgeon to become endangered.

Common sturgeons live in shallow seas around Europe and migrate up rivers to breed. They live on muddy or sandy bottoms and feed by stirring the sediment with their snouts and feeling for food with their sensitive barbels. At sea, their main foods include worms, crustaceans, shellfish, and some fish, such as sand eels. In rivers, they eat insects, worms, crustaceans, and shellfish. After spending a few years at sea, sturgeons swim up rivers to spawn in gravel-bottomed pools. Females lay 800,000 to 2,400,000 eggs.

Common sturgeons have disappeared from many rivers because the females are killed for their eggs before they are laid. They are extinct in once-important rivers, such as the Elbe, Rhine, and Vistula, and commercial fisheries for sturgeons have disappeared. Pollution of the rivers, the construction of dams, and other developments have also contributed to the sturgeon's rare status. Fish farms for common sturgeon have been established in France and elsewhere for producing caviar. These farms may help reduce the numbers of wild common sturgeons being killed, and it may be possible to release young sturgeons into the wild. Otherwise, the common sturgeon will survive only if it is strictly protected in its remaining breeding rivers.

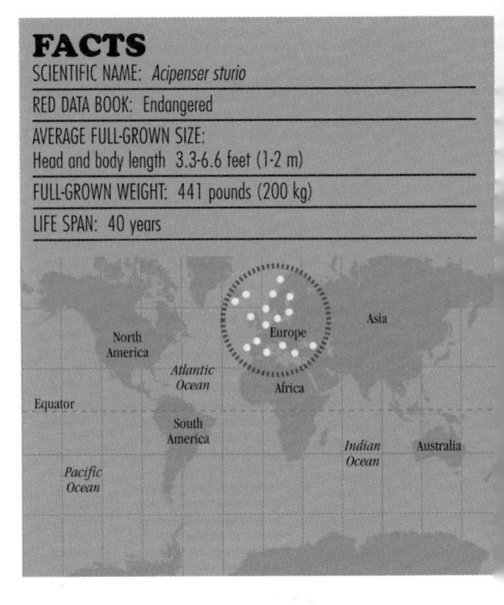

FACTS

SCIENTIFIC NAME: *Acipenser sturio*

RED DATA BOOK: Endangered

AVERAGE FULL-GROWN SIZE:
Head and body length 3.3-6.6 feet (1-2 m)

FULL-GROWN WEIGHT: 441 pounds (200 kg)

LIFE SPAN: 40 years

Common sturgeons use their sensitive barbels to locate food in sand and sediment.

BROAD SEA FAN

This close-up detail of a broad sea fan shows the polyps extended to trap microscopic food particles.

Sea fans are a type of coral. Instead of a heavy skeleton of lime like the true corals, sea fans and other soft corals have fragile, horny skeletons. They grow into delicate forms attached to rocks or true coral and look like plants. Each sea fan is a collection of individual animals called a colony. The broad sea fan lives in coastal waters of the eastern Atlantic, from Mauritania and perhaps the Gulf of Guinea to the British Isles, where it lives as far north as the western coast of Ireland and perhaps Scotland. It also lives in the western Mediterranean Sea. In different parts of the world, sea fans may be yellow, pink, or white.

Sea fans live in water deeper than 33 to 65 feet (10 to 20 m), but they sometimes come closer to the shore. They grow in shady areas where currents of water carry tiny floating animals past the branches of the sea fans. The animals are caught by stinging tentacles and digested. The broad sea fan grows slowly, adding only .394 inch (1 cm) per year. It reproduces by shedding tiny larvae into the sea. These larvae quickly settle near the parent and grow into adult sea fans.

Sea fans used to be popular souvenirs, especially in the United Kingdom. They are collected by divers, then dried and mounted on wooden blocks. Divers can quickly reduce a population of sea fans. This is dangerous for the species because it grows so slowly. Sea fans large enough to sell are over twenty years old. The threat in British waters has been reduced by warning divers that sea fans could become endangered, but there is nothing to stop a commercial organization from collecting and selling them in large quantities.

FACTS

SCIENTIFIC NAME: *Eunicella verrucosa*

RED DATA BOOK: Insufficiently known

AVERAGE FULL-GROWN SIZE:
Height 11.8 inches (30 cm)

FULL-GROWN WEIGHT: Not known

LIFE SPAN: Not known

BLACK CORAL

There are about 150 species of black corals. They live in all warm seas, and a few live in temperate regions. Black corals are collected for use in jewelry or for sale as souvenirs. They are not threatened with extinction, but they may become so rare that the trade in black coral will no longer be profitable.

Black corals live in deep water, usually between 98.4 and 360.9 feet (30 and 110 m), but some live as deep as 19,686 feet (6,000 m) and others in water as shallow as 3.3 feet (1 m). In the Galápagos Islands, black corals live in shallow water, growing in caves or under overhanging ledges, because the surrounding sea is too cold. Corals grow continuously, and Hawaiian black coral grows in height at a rate of 2.4 inches (6 cm) per year. The corals feed on tiny animals carried in the sea currents that are trapped and killed by the corals' stinging tentacles.

Black coral has been exported from the Persian Gulf to India for centuries. As well as its use in jewelry, black coral was prized for medicinal and magical properties. It is now important in the tourist trade, as well as for export, and it is becoming hard to find in many places, especially in the Caribbean. Divers collect the corals with an ax or hammer and float them to the surface with airbags.

Collecting black coral can provide an important source of income for people in tropical countries. The coral colonies in shallow water are in danger of disappearing, although they will survive in deeper water or in inaccessible places. Many attempts have been made to control the exploitation. In Hawaii, corals less than 47 inches (120 cm) cannot be collected. In some places, protection does not work because pirate collectors gather undersized corals and sell them cheaply. Marine sanctuaries will help preserve coral colonies if the areas can be properly policed.

Colonies of black coral may become restricted to waters that are too deep for divers to collect the corals easily.

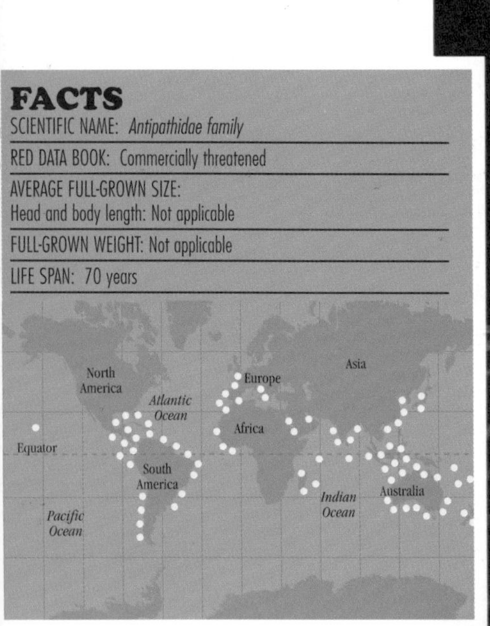

GIANT CLAM

FACTS

SCIENTIFIC NAME: *Tridacna gigas*

RED DATA BOOK: Vulnerable

AVERAGE FULL-GROWN SIZE:
Head and body length 51 inches (130 cm)

FULL-GROWN WEIGHT: 441 pounds (200 kg)

LIFE SPAN: 100 years

A giant clam in the waters off the Pacific island of New Caledonia.

The giant clam is the largest shellfish in the world. Although it can weigh up to 441 pounds (200 kg), only 121 to 143 pounds (55 to 65 kg) of that is living animal. The rest is the two huge shells. Giant clams live in the tropical waters of parts of the Pacific Ocean, from the Philippines to Micronesia and Australia. Other species of giant clam also live around the Indian Ocean.

Giant clams live only in shallow water on coral reefs, resting on sand or coral growths, sometimes in water so shallow that the shells are exposed at low tide. They do not live below 65.5 feet (20 m) in depth. Giant clams feed by filtering tiny animals and plants from the water with their gills. They get extra food from microscopic algae that live in their bodies. Special transparent organs in the flesh of the clam focus light onto the algae, which manufacture food by photosynthesis. The clam uses some of this food. Giant clams breed by releasing large numbers of tiny larvae into the sea. These float for several days, then settle on the seabed, form shells, and grow into clams.

Local people have always eaten the meat of giant clams, and they use the shells for washbasins and tools. The clams are now fished commercially for their meat, which the Japanese use in sushi cooking, and for their shells, which are used for decorations. They are so easy to collect that they soon become rare where they are fished, and they have disappeared from many places.

Giant clams are now protected in many places, but it is not always easy to enforce the laws. The Great Barrier Reef is fully protected, and aircraft patrol the area to watch for poachers. Experiments with farming giant clams are ongoing. They can be reared in special tanks and transferred to the sea when they are still small.

ATLANTIC HORSESHOE CRAB

Horseshoe crabs, such as this one photographed on a beach in Massachusetts, have lived in Earth's seas for hundreds of millions of years.

Like the coelacanths, horseshoe crabs are "living fossils." They have lived in the seas for hundreds of millions of years, but there are now only four species left. The American horseshoe crab lives on the east coast of North America, from Nova Scotia to the Yucatan Peninsula in southern Mexico.

Horseshoe crabs live on the seabed in sandy and muddy bays and estuaries. They have been found at depths of 807 feet (246 m), but most live between 16.5 to 20 feet (5 to 6 m). They feed on slow-moving or stationary animals, such as worms and shellfish, as well as on seaweeds. Breeding takes place at high tide on full and new moons in summer. Huge numbers of horseshoe crabs then emerge from the sea, with the males clinging to the females. The females dig a hole in the sand and lay hundreds or thousands of eggs. Five weeks later, the eggs hatch, and the larvae swim out to sea. The eggs and larvae are eaten by birds that visit the nesting beaches when they are migrating.

The population of American horseshoe crabs has declined over the last one hundred years. Because they eat clams, they are often killed by people who dig for clams. Horseshoe crabs are also collected to make fertilizer and for use as fishing bait. A more modern threat to their numbers is the use of horseshoe crabs in medical research. Once numbers have been reduced, it will take years for the population to recover. The solution is to manage the harvesting of horseshoe crabs so their numbers remain high. This will also help conserve the migrating birds that feed on them.

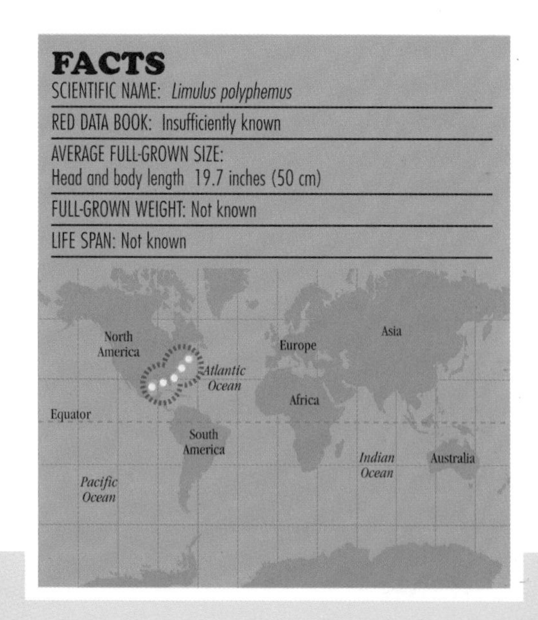

GLOSSARY

Aleuts — people living on the Aleutian Islands, west of Alaska.

aquatic — living or growing in water.

baleen — a hornlike substance formed into rows of plates that grow in the mouths of some whales. The baleen plates are used for straining food out of the water. Baleen used to be called whalebone.

barbels — fleshy organs near the mouth of a fish. Barbels are sensitive to taste and touch.

birds — warm-blooded animals that lay eggs and have feathers and wings.

burrows — holes dug into the ground or sand by animals to use as shelter.

carnivores — flesh-eating animals.

cetaceans — the group of mammals to which whales, dolphins, and porpoises belong.

clutch — a group of eggs in a nest, laid one at a time.

colonies — groups of plants or animals of the same kind living together.

compete — to strive against another in order to win something or to reach a goal.

conservation — the act of preserving plants, animals, or other resources from extinction.

copepods — small crustaceans with many pairs of tiny, oarlike legs for swimming. Copepods can be freshwater or marine.

crustacean — an invertebrate animal with a segmented body and an external skeleton or shell, e.g., shrimp, crab.

decimate — to ruin or destroy.

decline — to become weaker or move downward.

den — the dwelling or shelter of a wild animal.

dorsal fin — the fin on the back of a fish or cetacean.

El Niño — a warm ocean current that flows periodically around the western coast of South America, changing the overall climate and raising the water temperature.

endangered — in danger of dying out completely, or becoming extinct.

evolve — to change shape or develop gradually over a long period of time.

extinct — animals and plants that have died out.

fish — an aquatic vertebrate animal with gills for breathing and fins for swimming.

food chain — an arrangement of animals and plants in which each species is a food source for the next higher species in the series.

fossil — the mark or remains of a prehistoric animal or plant preserved in rock.

gills — breathing organs in the head of a fish, amphibian, or aquatic invertebrate animal. Gills extract oxygen from the water and release carbon dioxide from the body into the water. They are the equivalent of the lungs of a land animal.

habitat — the natural home of an animal or plant.

Inuit — the people who live in Greenland and Arctic America. They are sometimes known as Eskimo.

ivory — the hard, white substance of the tusks of elephants, walruses, narwhals, and other animals.

krill — shrimplike crustaceans that live in swarms and are eaten as food by whales and other animals.

lagoon — a shallow lake that opens into a sea or a river.

landlocked — surrounded or nearly surrounded by land.

larva — in the life cycle of insects, amphibians, or fish, the stage that comes after the egg but before full development; e.g., a caterpillar is the larva of a butterfly or moth.

mammals — warm-blooded animals, including whales and dolphins, that feed their young on mother's milk.

marine — living or growing in the sea.

marsh — waterlogged ground.

migration — the movement of animals from one environment to another, often over long distances, and usually following certain routes on a seasonal basis.

native — animals or plants that originate or occur naturally in a particular place.

photosynthesis — the process by which the energy of sunlight is converted by plants, using chlorophyll and carbon dioxide.

poaching — the illegal catching of animals, usually for profit.

pollution — the gas, smoke, trash, and other harmful substances that ruin our environment.

polyp — an animal, such as a sea anemone, with a hollow body and tentacles.

predator — an animal that feeds by catching and eating other animals.

pup — a young animal.

reserves — areas of land set aside for the protection of wildlife.

rorqual — a large whale with grooves under its chin and breast.

salt marsh — an area of flat land near the sea that floods at high tide.

sanctuary — a place or geographical area where animals are protected from hunters or other harmful outside influences.

school — a group of fish or cetaceans living together.

sirenians — a group of plant-eating, aquatic mammals that includes the dugong and manatee.

snout — protruding nose and jaws of an animal.

spawning ground — an area where large numbers of fish lay their eggs.

specimen — a single object, plant, or animal used for study.

subantarctic — the region bordering the Antarctic.

subsistence hunters — people who hunt animals for their own use as food or materials, not for sale.

tail flukes — the flat end of a whale's tail that propels it through the water.

temperate — the parts of Earth with a climate between the heat of the tropics and the cold of the polar regions.

trawl nets — nets that are pulled through the water to catch fish and other animals.

tropical — of or relating to the warm, humid area of Earth near the equator; the area of Earth that lies between the Tropic of Cancer and the Tropic of Capricorn.

vertebrate — an animal that has a backbone.

viable — large enough (as in a species population) to continue breeding and existing.

whale songs — the calls of male humpback whales during the breeding season.

MORE BOOKS TO READ

All Wild Creatures Welcome: The Story of a Wildlife Rehabilitation Center. Patricia Curtis (Lodestar)
The Californian Wildlife Region. V. Brown and G. Lawrence (Naturegraph)
Close to Extinction. John Burton (Watts)
Conservation Directory. (National Wildlife Federation)
Conservation from A to Z. I. Green (Oddo)
Discovering Birds of Prey. Mike Thomas and Eric Soothill (Watts)
Discovering Endangered Species (Nature Discovery Library). Nancy Field and Sally Machlas (Dog Eared Publications)
Ecology Basics. Lawrence Stevens (Prentice Hall)
Endangered Animals. John B. Wexo (Creative Education)
Endangered Forest Animals. Dave Taylor (Crabtree)
Endangered Grassland Animals. Dave Taylor (Crabtree)
Endangered Mountain Animals. Dave Taylor (Crabtree)
Endangered Species. Don Lynch (Grace Dangberg Foundation)
Endangered Species Means There's Still Time. (U.S. Government Printing Office, Washington, D.C.)
Endangered Wetland Animals. Dave Taylor (Crabtree)
Endangered Wildlife. M. Banks (Rourke)
Fifty Simple Things Kids Can Do to Save the Earth. Earthworks Group (Andrews and McMeel)
Heroes of Conservation. C. B. Squire (Fleet)
In Peril (4 volumes). Barbara J. Behm and Jean-Christophe Balouet (Gareth Stevens)
Lost Wide Worlds. Robert M. McClung (William Morrow)
Macmillan Children's Guide to Endangered Animals. Roger Few (Macmillan)
Meant to Be Wild. Jan DeBlieu (Fulcrum)
Mountain Gorillas in Danger. Rita Ritchie (Gareth Stevens)
National Wildlife Federation's Book of Endangered Species. Earthbooks, Inc. Staff (Earthbooks, Inc.)
Project Panda Watch. Miriam Schlein (Atheneum)
Save the Earth. Betty Miles (Knopf)
Saving Animals: The World Wildlife Book of Conservation. Bernard Stonehouse (Macmillan)
Why Are Animals Endangered? Isaac Asimov (Gareth Stevens)
Wildlife Alert. Gene S. Stuart (National Geographic)
Wildlife of Cactus and Canyon Country. Marjorie Dunmire (Pegasus)
Wildlife of the Northern Rocky Mountains. William Baker (Naturegraph)

VIDEOS

African Wildlife. (National Geographic)
The Amazing Marsupials. (National Geographic)
Animals Are Beautiful People. Jamie Uys (Pro Footage Library: America's Wildlife)
How to Save Planet Earth. (Pro Footage Library: America's Wildlife)
Predators of the Wild. (Time Warner Entertainment)
Wildlife of Alaska. (Pro Footage Library: America's Wildlife)

INDEX

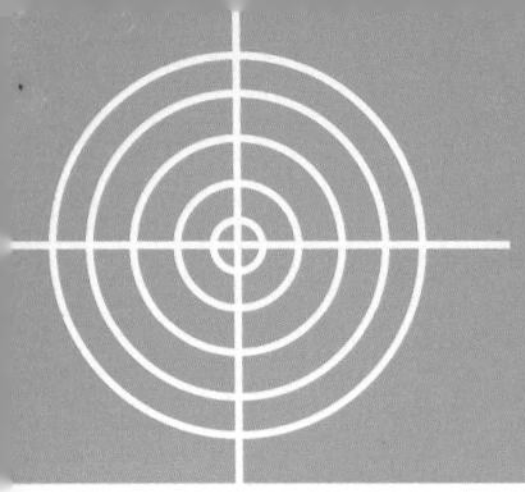

PICTURE CREDITS

Page 9, courtesy of Richard Ostfeld
Page 10, 49, Michael Brooke
Page 11, © Earthviews/FLPA
Page 12, © Mark Newman/FLPA
Page 13, Tom & Pam Gardner/FLPA
Page 14, © Christiana Carvalho/FLPA
Page 15, © Domenico/Panda Photo/FLPA
Page 16, © D. Cavagnaro/Panda Photo/FLPA
Page 17, © Rodger Jackman/OSF
Page 18, Dan Guravich/WWF
Page 19, WWF Canada/© WWF
Page 20, Peter Best/WWF
Page 21, James Watt/Panda Photo/© WWF
Page 22, © G. Williamson/Earthviews/FLPA
Page 23, © Earthviews/FLPA
Page 24, K. Balcomb/WWF
Page 25, 50, 51, James Watt/Panda Photo/WWF
Page 26, © R. Pitman/Earthviews/FLPA
Page 27, G.W. Miller/© WWF
Page 28, Gregory Silber
Page 29, John Woestendiel/© Earthviews/FLPA
Page 30, Dr P.K. Anderson/WWF
Page 31, Henry Ausloos/BIOS/WWF
Page 32, Gerald Cubitt/WWF
Page 33, © Don Roberson/BirdLife International
Page 34, Peter Harrison/VIREO/© CCD Productions
Page 35, 36, BirdLife International
Page 37, Don Hadden/© VIREO
Page 38, © Philip Perry/FLPA
Page 39, © Richard W. Beales/Seaphot Limited: Planet Earth Pictures
Page 40, H. Witt/WWF
Page 41, OSF
Page 42, Michel Gunther/WWF
Page 43, Mauri Rautkari/WWF
Page 44, Arquivo Projeto Tamar/WWF
Page 45, P. Pritchard/WWF
Page 46, © Silvestris/FLPA
Page 47, © M. Gore/FLPA
Page 48, © Tom & Pam Gardner/FLPA
Page 52, © Howard Hall/OSF
Page 53, Peter Scoones/© Planet Earth Pictures
Page 54, © Frank W. Lane/FLPA
Page 55, 58, © D.P. Wilson/FLPA
Page 56, © Panda Photo/S. Navarrini/FLPA
Page 57, R. Seitre/BIOS/WWF

FLPA= Frank Lane Picture Agency
WWF = World Wide Fund for Nature
OSF = Oxford Scientific Films
VIREO = Visual Resources for Ornithology